SOCIAL JUSTICE IN EDUCATION

SOCIAL JUSTICE IN EDUCATION

AN INTRODUCTION

BARRY L. BULL

SOCIAL JUSTICE IN EDUCATION
Copyright © Barry L. Bull, 2008.

First published in 2008 by
PALGRAVE MACMILLAN®
in the United States—a division of St. Martin's Press LLC,
175 Fifth Avenue, New York, NY 10010.

Where this book is distributed in the UK, Europe and the rest of the world,
this is by Palgrave Macmillan, a division of Macmillan Publishers Limited,
registered in England, company number 785998, of Houndmills,
Basingstoke, Hampshire RG21 6XS.

Palgrave Macmillan is the global academic imprint of the above companies
and has companies and representatives throughout the world.

Palgrave® and Macmillan® are registered trademarks in the United States,
the United Kingdom, Europe and other countries.

ISBN-13: 978–0–230–60650–0
ISBN-10: 0–230–60650–4

Library of Congress Cataloging-in-Publication Data

Bull, Barry L.
    Social justice in education : an introduction / Barry L. Bull.
        p. cm.
    Includes bibliographical references.
    ISBN 0–230–60650–4
      1. Education—United States—Philosophy. 2. Social justice—
United States. I. Title.

LA210.B793 2008
370.11′5—dc22   2008004246

A catalogue record of the book is available from the British Library.

Design by Newgen Imaging Systems (P) Ltd., Chennai, India.

First edition: October 2008

10 9 8 7 6 5 4 3 2 1

Printed in the United States of America.

For
Irene Ruderman Bull

CONTENTS

PREFACE

Social justice has become a revitalized theme in much of the educational writing, research, and policy making in the past decade or so, as concern over a variety of issues has emerged in the field about the fairness of the education system to marginalized students and even to students in general. One such issue is the gap in academic achievement between, on the one hand, majority students and racial and ethnic minority students and, on the other, students from socioeconomically disadvantaged families and other students. But many other issues have received attention as well, such as a concern about the fairness of the increased focus on uniform standards and accountability for all students and schools required under the federal No Child Left Behind Act and state education policies; the inequalities in high school graduation rates among different student populations; indications of intensified racial and economic segregation in schools, particularly in urban school districts; the shift from school funding systems aimed at providing an equitable education to those aimed at providing an adequate education; and the lack of progress in equalizing student access to college as, for example, the federal government, states, and institutions increasingly substitute financial aid based on academic merit for that based on financial need and hold down the value of need-based aid. There is good reason why social justice has become a major theme of recent thought about American public education. After all, the public school system is one of this country's major social institutions, one that, like few others, significantly, continuously, and inescapably affects all citizens, whether as students, parents, employers, workers, or taxpayers. If we do not expect this institution to treat citizens fairly in these various roles, the hope that we will be able to achieve anything like a recognizably just society is significantly undermined.

Although attention to social justice continues to grow in the field, at least two difficulties with the treatment of the subject have emerged. First, there has been a simultaneous growth in concern about cultural diversity in schools so that prior assumptions about the mechanisms to deliver social justice that were based on uniformity in what children and their families deserve from an education—access to the same curricula, instruction aimed at the same goals for academic achievement, mechanisms that provide the same opportunities for parent involvement, and so on—have increasingly come under fire. Therefore, prior conceptions of the meaning of social justice itself have come to seem increasingly inadequate, and, for some, social justice appears to be a questionable social value. Second, in much of this recent work, social justice is often narrowly construed, as dealing with the fair treatment of only one segment of the population (for instance, children or employers) or as involving a single value (for instance, equal opportunity or economic competitiveness). Therefore, the proposals that result from such work often do not speak to their wider consequences for achieving the complex and many-faceted ideal that social justice represents. This limited focus is particularly troubling when comprehensive policies—for accountability or funding the system, for example—are at issue precisely because such policies have potentially significant consequences for all of the system's clients, functions, and effects. The wide ramifications of these policies make social justice, with its societywide focus, the premier normative criterion for their evaluation, and those ramifications also make it imperative that the fullest meaning of social justice be brought to bear in this evaluation. For these reasons, the time is right for a systematic contemporary account of social justice in education. This book articulates and argues for a set of principles that encompass an account of the broadly inclusive meaning of social justice for schools and then considers its implications for several of the most general types of policies in American education—namely, curriculum content, instructional procedures, education standards, civic education, control of schools, and school finance.

At the outset, it is important to be clear about the ways in which this book's purview is intentionally incomplete with respect to the broad normative dimensions of education. First, this is a book about education in the United States. As will be explained in text, it begins with Americans' intuitions and settled convictions about how their schools are to operate and what they are to achieve. Therefore, it does not pretend to consider the equally important, and in some cases even more urgent, issue of what social justice might imply for education in other societies where citizens may have different intuitions about these matters. Of course, to the extent that citizens of other countries share some or all of these intuitions, the book may be helpful in thinking through their ramifications for the framework, policies, or conduct of education in those societies. This is not to say that the issue of international justice in education is unimportant, that it cannot be given thoughtful and systematic treatment (see, e.g., Nussbaum 2006), or that what Americans think is not relevant to this larger topic, but only that it is not the explicit subject of this book's methodology or content.

Second, this is a book about education and, as such, it does not include discussions of what social justice in the United States might be on a whole variety of other important subjects, such as health care, income distribution, or fighting crime. Because these subjects may interact in significant ways with education, the ramifications of these other dimensions of social justice may, therefore, imply normative constraints on the application or fulfillment of the principles of social justice in education (e.g., by making the priorities for the use of public resources more complex), or those principles might, in turn, constrain the policies adopted to achieve justice in other social arenas (e.g., by adding to the considerations that apply to the treatment of juvenile offenders). In this way, this book can be understood as a tentative and incomplete contribution to a wider project that seeks principles of justice for American social institutions more generally.[1] Obviously, these other arenas of justice and the wider project are important, but that Americans' moral intuitions about education seem to be more harmonious and settled than they are on these other subjects makes education a reasonable place to start. At any rate, it seems unnecessary and probably unwise to wait for the development of an acceptable and all-inclusive theory of social justice for the full array of American institutions before acting in good faith to achieve that goal, however incompletely, in our schools.

Third, social justice is not the only public normative consideration relevant to education. In particular, the personal and professional ethics of teachers, administrators, counselors, and other school employees are only partially determined by social justice. Thus, the normatively appropriate relationships between teachers and students undoubtedly involve issues of, for example, personal caring about which social justice may have little to say (see, e.g., Noddings 2003). Or the organization of schools may raise questions about whether and how to establish healthy institutional communities (see, e.g., Strike 2004) that may also be beyond the scope of social justice. To be sure, these other normative considerations have sometimes been interpreted as competitors with social justice. However, in my judgment at least, it is just as plausible to understand these values as at least partially independent of and complementary to social justice rather than as largely antithetical to it. Thus, this book is incomplete in that it omits a consideration of such other important moral and public values. Even so, social justice is a significant public value, particularly in the United States, and it expresses substantial normative expectations about the public schools and American citizens' responsibilities for them. Identifying and acting on social justice in schools are, therefore, at least part of what public morality requires of and in these institutions.

Fourth, this book is intended to provide a somewhat abstract and relatively brief introduction to the subject of social justice in education. On the one hand, this implies that the primary focus is on clarifying the meaning of and providing arguments for the general principles at work in Americans' conception of social justice in schools. On the other, it implies that the specific policies that the book analyzes have been selected largely because of their capacity to bring out aspects of that meaning in addition to what a purely theoretical consideration can make apparent. Thus, there are many implications of social justice that are not considered in this book—on general topics, such as parent and student choice of schools or the relationship between schools and higher education, and on specific policy options, such as school vouchers and the appropriate form and availability of the college preparatory curriculum in high schools. Moreover, there are important practical matters about the implementation of the policies that are discussed—such as the details of the school curriculum, of the treatment of children with disabilities, or of school funding formulas—with which this book does not deal. Yet, I believe, social justice is a significant consideration in formulating and evaluating these and many other policies and practices that usually requires careful attention to the context of particular communities and schools in addition to the application of the principles themselves. In effect, then, there are many applications and implications of the principles of social justice in education that this book leaves to the energy, imagination, and intelligence of the concerned reader.

Finally, this book is intended primarily to articulate, explain, and lay out the grounds for a particular conception of social justice in education. As such, it does not attempt to formulate a systematic response to all of the many alternative philosophical conceptions of social justice, even those that concentrate on education. It does include responses to some elements in the current education and philosophical literature when those responses are useful in clarifying the meaning or justification of this conception, but constructing a systematic defense of these ideas against all or even many of the alternatives, although of great intellectual and moral importance, is not this book's primary purpose.

I would like to acknowledge that the content of several chapters draws on work that I published previously. Chapters two and three are adapted from "A political theory of social justice in American schools," in *To what ends and by what means?: The social justice implications of contemporary school finance theory and policy,* edited by Gloria M. Rodriguez and R. Anthony Rolle (New York: Routledge, 2007), 9–34. Chapter four borrows from "Is standards-based school reform consistent with schooling for personal liberty?" *Studies in Philosophy and Education* 25, no. 1 & 2 (2006) and from "National standards in local context: A philosophical and policy analysis," in *Educational leadership: Policy dimensions in the 21st century,* edited by Bruce A. Jones (Stamford, CT: Ablex Publishing, 2000), 107–121. Chapter five uses ideas developed in "Can civic and moral education be distinguished?" in *Moral and civic learning in the United States,* edited by Donald Warren and John Patrick (New York: Palgrave/Macmillan, 2006), 21–31; these ideas are developed further in "A Politically Liberal Conception of Civic Education," *Studies in Philosophy and Education* (In press). Chapters six and seven draw on "Political philosophy and the balance between central and local control of schools," in *Balancing local control and state responsibility for K-12 education: 2000 yearbook of the American Educational Finance Association,* edited by Neil Theobald and Betty Malen (Larchmont, NY: Eye on Education, 2000), 21–46.

This book has already benefited in innumerable ways from the critical intelligence of interested colleagues around the country and the world. I want to thank the doctoral students at Indiana University for the opportunity to discuss these ideas as they were being formulated and refined over many years and especially the members of the student discussion group in philosophy of education who read, criticized, and made useful suggestions on an early draft of the entire text—Chris Hanks, Dini Metro-Roland, Joe Ramsey, and Lyndsay Spear. I am especially grateful to my colleagues in the field for their invitations to write about these ideas over the years as contributions to a variety of scholarly projects—most notably, to Bob Arnove, Virgie Chattergy, Royal Fruehling, John Goodlad, Bruce Jones, Betty Malen, Martha McCarthy, John Patrick, William Reese, Gloria Rodriquez, Anthony Rolle, David Smith, Kenneth Strike, Margaret Sutton, Neil Theobald, and Donald Warren. Of particular value in the refinement of this book was Kenneth Strike's reading of and critical commentary on several draft chapters. I am also indebted to Donald Warren for reading a draft of this book and for giving me his thoughtful advice about its audience and publication. Finally, the many scholars who have attended the biennial meetings of the International Network of Philosophers of Education have given me a chance to clarify and sharpen these ideas for an international audience. Of course, any errors or inelegancies that remain reflect my own inability to benefit from their wise critique and sound advice.

CHAPTER ONE

THE NATURE AND IMPORTANCE OF SOCIAL JUSTICE IN EDUCATION

Jamesville is a small Midwestern town that has experienced in the past few years an influx of Latino heritage families who have migrated to work in its burgeoning meatpacking industry. Carrington Middle School serves all of Jamesville's students in the sixth, seventh, and eighth grades. In the past, Carrington students performed well on the state's mandatory standardized tests of English and mathematics, and the community took pride in the school's ranking among the top 25 percent of middle schools in the state. Three years ago, however, the school's test scores began to dip. At first, school officials attributed the decrease to normal year-to-year variation, but the scores continued a downward trend in the two succeeding years, so that currently Carrington test scores rank at about the average for the state. Even worse, the annual state reports on the test scores required by the federal No Child Left Behind Act labeled the school as failing to make Adequate Yearly Progress (AYP) during those years. That legislation and the state regulations that implement it require every school in the state to demonstrate a specified improvement each year in the proportion of students who score at or above the minimum set by the state. Not only must the school's entire student body show such progress, but also the school must demonstrate similar progress for each of several distinct categories of students, including those who qualify for the federally subsidized lunch program and those who have limited proficiency in English. The state test reports clearly demonstrate that the school's failure to meet AYP is largely the result of the performance of its recently arrived Latino students, whose numbers have grown to nearly 20 percent of the school's student population. On the basis of its proportion of children from low-income families, Carrington qualifies to receive federal Title I funds to supplement the instruction of students with low academic performance. Therefore, the state has begun the federally mandated process of imposing sanctions on the school for its failure to meet AYP.

As might be expected, the anxiety in Jamesville over this development has been widespread. The school board and district administrators complain that the state's testing standards do not adequately take into account the rapid change in the town's demographics. Middle school teachers express similar concerns but also think that the district and the state have not provided them with sufficient resources and assistance to teach Latino students effectively and, pointing to a similar but less dramatic decline in elementary students' test scores, believe that the district's elementary schools are not preparing these students adequately for middle school work. Many Anglo parents, while expressing some sympathy about the unfair labeling of the middle school, are convinced that the test scores are evidence that the schools are beginning to lose focus on their traditional academic mission. Latino parents have been less vocal in the public debate, but their community leaders have quietly expressed the worry that Latino students and their parents are being unfairly if surreptitiously blamed for the test score decline, which, they believe, reveals the schools' systematic neglect of Latino students.

These initial reactions of Jamesville's citizens reflect an understandably self-interested response. First, everyone seems to want to deflect the blame away from themselves and toward another segment of the community or the educational system. School officials blame the state; middle school teachers blame the state, the district, and elementary schools; parents blame the school. Second, they interpret the problem as revealing an inadequacy in the treatment of themselves or those they care most about. School officials believe that they are the victim of a state accountability system that is insensitive to the community's circumstances; teachers do not believe themselves to have the educational resources to meet the state's expectations; Anglo parents worry that their own children are being deprived of the education that they had come to expect; Latino parents think their children are being denied the instructional support that will enable them to succeed. Finally, no one anticipates corrective actions that might require him or her to revise their own values or behavior. School officials want the state to change its policies and procedures; middle school teachers want more funding from the state and district; parents want the teachers to modify their attitudes toward and their instruction of their children.

Self-interest, however, is not necessarily irrelevant to forming a reasonable judgment about Carrington Middle School and to devising an appropriate course of action. In a sense, all interests are self-interests in that they are interests held by individuals, but they are not necessarily selfish interests. After all, at least some of the self-interests on display in Jamesville may prove to be of social importance that is of concern to others beyond those who are immediately affected. Thus, others in the community or the state may have good reason to be concerned about the alleged discrimination that initially is expressed as arising from the self-interests of Latino parents, for example. Other self-interests that have not been directly expressed so far in this situation may also be pertinent to reaching a judgment about it—the interests, for example, of the other citizens of the state and the nation that are affected by the performance of the Jamesville's children and the interests of the children themselves that are not encompassed by their parents' or other adults' concerns. Thus, the businesses that will employ those children as adults are likely to have a long-term stake in these children's school achievement, and the children themselves may have interests in their own academic progress connected with aspirations for their lives beyond what the adults in their community recognize and value. Moreover, the interests at issue are not exhausted by the consequences of these children's school performance, that is, by what they come to know and to do as a result of their education. People may also have an interest in the appropriateness of the means by which those results of education are achieved, in, for instance, whether the instructional process and the accountability system are humane, whether access to instruction is appropriately distributed among the town's children, whether the consequences of the accountability system fall unfairly on particular segments of the community, and whether the costs of supplying those means fall on those who have a responsibility to pay for them.

This context of multiple and interlocking interests is typical of school controversies of all kinds. After all, public schools are the focus of sometimes similar but often competing aspirations of many different individuals and groups in our society, most immediately children, parents, teachers, and school administrators but also ethnic and religious groups, government officials, the business community, and taxpayers. Furthermore, the various members of these groups themselves frequently disagree in their perceptions of what is at stake in these controversies. Clearly, parents often have different expectations of schools, with some emphasizing schools that show their children kindness and caring, some wanting schools to support and reinforce the families' religious and cultural values, and others seeking schools that can supply economic or intellectual opportunities for their children that the families, for one reason or another, do not or cannot provide. Moreover, for example, members of the business community have diverse expectations, with some placing the emphasis on the schools' graduates filling their immediate demand for labor, others on the schools' effect on their ability to recruit and retain appropriately skilled parents as workers, and yet others on the schools' influence on the nation's economic development in the long term.

These various interests imply that the issues of Jamesville's schools are, on the one hand, essentially political. In general, politics speaks to the ends for which and the methods by which a society is governed. The issues in Jamesville schools arise in part from the actions taken by governments at a variety of different levels—local distribution of resources, state imposition of student performance standards, and federal requirements for school accountability. Many of the responses to those issues take the form of concerns about the appropriateness of the actions of those governments—among other things, whether local resources are being distributed effectively, whether state standards are being enforced fairly, and whether the federal scheme of school accountability is sufficiently sensitive to local contexts. Government involvement is thus a fact of public schooling generally and in this particular instance. But beyond this, public schooling has been one of the most basic government functions in this country since the mid-nineteenth century. It is, of course, imaginable that governments could stay out of the schooling business in the same way that it largely (although not entirely) leaves manufacturing and consumption decisions about shampoo and television sets to individual initiative and taste, but even the most radical proponents of school privatization do not advocate complete government withdrawal from school regulation and funding. In other words, there is wide agreement that governments should in some way provide for and oversee schools. In this way, school controversies, such as the one in Jamesville, seem to be permanent fixtures of American political life.

At the same time, these issues are essentially moral; that is, they concern what should be done in and by schools in light of the rights and well-being of those affected. Of course, these issues also reflect, in part, disagreements over matters of fact. For example, the Latino community's concerns are based on their beliefs that Latino children are receiving educational services that do not enable them to succeed in school, and Anglo parents believe that schools are becoming ineffective in teaching their children the traditional academic curriculum. Resolving these issues clearly does require ascertaining whether these factual allegations are true. But such a factual investigation, on its own, will not resolve the issues because the Latino and Anglo parents also believe that their children have a right to instruction that is effective in the particular ways that they care about. Thus, the Anglo parents' allegations of fact might turn out to be true, but a change in instructional focus to accommodate the needs of Latino students more adequately might be morally justified because such a change enables the schools to achieve results that are of higher moral priority than individual or aggregate academic achievement. In other words, parents' and other citizens' beliefs include both descriptive and moral elements, and a resolution of the issues that they raise requires an evaluation of both.

The Relevance of Social Justice

One helpful way of understanding these complex controversies that simultaneously involve political, moral, and factual disagreements was suggested by the American political philosopher John Rawls (1993, 1999b). He noted that a hallmark of societies such as ours is that they enable their citizens to maintain what he called their own conceptions of the good. These conceptions involve beliefs about what a life worth living consists in—the aims that one should pursue and the obligations that one has in pursuing them. They also frequently include what Rawls termed metaphysical commitments—beliefs about the fundamental matters that make these aims and obligations justified, for example, beliefs about the divine, the order of the universe, and humans' role in that order.

However, several explanations and clarifications of this idea of conceptions of the good are in order. First, the beliefs that constitute those conceptions are not exclusively or even primarily expressed in such grandiose and abstract terms as this general description might lead one to suppose. Rather, they include ordinary commitments such as one's beliefs about decent work, conscientious relationships, and fulfilling activities, among a wide range of other things. Second, conceptions of the good are rarely rigorously consistent or complete.

Instead, they often include beliefs that contradict one another, and they often neglect many matters about which one has no firm conviction. Third, even though they are personal configurations of belief, attachment, and involvement, conceptions of the good are rarely one's own original inventions. Rather, they most often include beliefs that are drawn from the array of traditions in one's society that are shared by the members of various religious, cultural, geographic, or socioeconomic groups. Fourth, while they include beliefs about one's own actions, conceptions of the good also involve expectations about others' behavior. In part, they include such expectations of others because others' behavior can restrict the latitude that one has to act in the ways that one most prefers. But also, others' behavior can violate one's commitments even if one is not directly affected by such behavior. For example, a person can care about the extent of violence in one's society because one is worried about its direct effects on one's own life or the lives of one's family members. But even if the level of violence does not directly affect oneself or one's loved ones, one can be concerned about it because of a belief that all people have a right to live securely. Thus, one's conception of the good can determine one's response to actions performed by others in far distant locations or with whom one rarely or never interacts. Finally, conceptions of the good are not usually static or permanent. Rather the beliefs they include often change with our own experience and with our coming to learn about the conceptions and the experiences of others. In fact, many hold conceptions in which such change is regarded as an inherent and valued aspect of living.

In light of this discussion, the factual and moral disagreements about schools and much else in Jamesville can be seen to result from the differential conceptions of the good held by the citizens of that community, its state, and the nation. Clearly, such conceptions can involve commitments to altogether different moral standards or, perhaps more frequently, different rankings of the importance of particular moral standards that are held in common, and we see clear evidence of such variation in parents' priorities for their children's education in Jamesville. It is not as if, for example, Latino parents do not value academic excellence in children's education and Anglo parents do not value fairness in the distribution of instructional resources. However, these parents' different conceptions of the good combined with their different experiences and circumstances lead them to place different priorities on these values, at least in the conditions that prevail in Jamesville. Beyond this, moreover, different conceptions of the good in conjunction with different experiences can lead to different perceptions and interpretations of empirical matters. Thus, for example, what some parents see as a reduced emphasis on rigorous academics can be understood by others as an effort to be more inclusive of the interests of previously neglected children. Of course, the events at Carrington undoubtedly include both of these things as well as much else, but the differential values and circumstances of some parents lead them to emphasize one interpretation of what is happening in the schools in preference to other possibilities.

At first glance, the politics that arise from these different conceptions of the good seems to be as hopelessly mired in disagreement as the facts and values of the situation. After all, politics can be understood as simply a way of each citizen's urging on others the interests that his or her conception of the good and perceptions entail, and therefore the political positions that citizens take seem to be as inevitably varied and as often opposed as their personal conceptions of the good are. From this perspective, then, politics seems to be at best citizens' effort to form coalitions with the like-minded to enforce their will on those who disagree. Indeed, some political theorists have suggested that such interest-group struggle is precisely what politics in a democracy amounts to (Dahl 1971).

John Rawls (1993), however, sees another possibility for liberal democratic politics. He argues that those who hold different conceptions of the good can seek what he calls an overlapping consensus about the general moral principles that should govern their political association. For Rawls, in other words, despite wide disagreement about the good life, a consensus about the public rules that govern political relationships, rights, and obligations may still be possible. Such consensus can arise not only in part from the explicit values and beliefs that people just happen to share but also in part from their deliberations about the conditions that seem necessary or at least useful in enabling people to live in ways that they regard to be worthwhile. Indeed, he argues that such a consensus is implicit in the very idea of permitting citizens to develop, maintain, and pursue the conceptions of the good that they find reasonable and personally satisfying. Thus, a wide toleration of and, indeed, respect for others' conceptions of the good with which one disagrees can be seen to be a condition of one's having a secure opportunity to live a life that one finds to be worthwhile. In this way, these underlying values may form part of the overlapping consensus in a widely diverse society, whether they happen initially to be shared by all the citizens of that society. I have much more to say about such an overlapping consensus about schools in the chapters that follow, but for now it is important to note that those who believe differently about the goods that they pursue in life can nevertheless agree, for example, that all have a right to pursue the goods that seem best to them, especially if they reason carefully about the various things that they value initially, how those values might be reasonably adjusted when they come into conflict, and what is generally supportive of and useful to their realizing those modified but still varying values.

The attraction of the idea of an overlapping consensus is that it permits us to establish a political order based on a foundation of moral principles that are acceptable to citizens with a wide plurality of personal belief and practice and that secure the conditions for maintaining and protecting that pluralism. Indeed, such moral principles, if they prove possible, would provide a conception of justice for such a society—a conception that treats the adherents to various conceptions of the good fairly by, on the one hand, adjusting its public principles to what the members of the society actually can in good conscience accept and, on the other, respecting and facilitating the various ways of life that those members of society deem best for themselves. In short, an overlapping consensus provides an account of social justice for a society characterized by a wide variety of personal conviction and preferred activity, namely, a society like our own.

Now, the relevance of such an abstract idea as an overlapping consensus to the concrete disagreements over the schools is Jamesville, or any other community for that matter, may not be immediately apparent, and in large part that relevance for the conduct of schooling is what this book aims to establish. Of course, the details of any specific application of this idea to the situation in Jamesville would require an account of what principles an overlapping consensus about education in America actually includes. I could simply assert those principles here and go on to show how they apply to the situation in Jamesville. However, it is more intellectually honest and ultimately more persuasive to provide arguments for those principles before reaching any conclusions about what follows from them. However, I can at least suggest in a general way how principles of social justice when understood as an overlapping political consensus might be helpful in assessing and resolving at least some aspects of the situation.

To continue the example of the different perspectives of Latino and Anglo parents on their children's education, we have already noted that it is likely that both parties hold similar enough values to enable them to understand and perhaps to appreciate one another's concerns. That is, if Anglo parents were to perceive Carrington's previous curriculum and instructional practices as inhibiting the realization of their aspirations for their children, it is likely that they would react in much the same way that many Latino parents have. Similarly, if Latino parents perceived those practices as advancing their expectations for their children, they would be as concerned about the changes that have taken place at Carrington as many Anglo parents are. Thus, it is clear that these two groups of parents share at least some values despite their differences of culture, language, and experience. In this case, they at the very least share a value that parents' aspirations should have an influence on their children's schooling, and probably they share much more about the specific content of that schooling. What an overlapping consensus seeks to identify and describe is the immediate values that Latino and Anglo parents happen to share, if any, for their children's education and, more importantly, the underlying values that must be affirmed if parents of any ethnicity are to have a right to expect that their aspirations for their children should influence their children's education. In light of these shared and underlying values, we may be able to interpret the situation in Jamesville in a way about which all these parents can in principle agree. And that more fundamental interpretation of the situation at least raises the possibility that at least some of the differences of opinion between the two groups of parents can be resolved in ways that both can understand and accept. These resolutions do not necessarily mean that all parties will be completely satisfied with the results. Indeed, such an outcome is unlikely, given the many remaining differences between them, but justice promises not that everyone will get his or her own way but only that the results will be fair to all.

Of course, arriving at an overlapping consensus about education is more complicated than this example might suggest. After all, parents are not the only interested parties in this situation, for future employers, teachers, government officials, and the children themselves have a stake in the outcome. In determining the legitimate role of parents' aspirations in their children's education, these other interests will have to be consulted. Thus, a genuinely overlapping consensus must take account of the interests and perspectives of all these parties and undoubtedly many others. But if such a consensus turns out to be possible, it holds out the prospect of a political resolution for at least some of the debates about American education that is more principled and morally justifiable than simply resolving them by means of a politics of naked interest-group power.

The Purpose and Structure of this Book

In this book, I explore the possibility for and the meaning of an overlapping consensus about the public principles that apply to the education of children in our society. Such principles can help us to understand and to resolve, at least partially, disagreements such as those I sketched in Jamesville. In chapters two and three, I derive the basic principles of social justice for education based on an effort to reflect on and to render consistent with one another four widely held public purposes for education in our society—personal liberty, democracy, equality of opportunity, and economic growth. Admittedly, the principles that emerge from these arguments are highly abstract and, thus, need further clarification and explanation so that they can be applied to the disagreements that emerge about American schools. As a result, in each subsequent chapter, I elaborate on the meaning of one of the principles and apply it to a current controversy over public policy in education. In chapter four, I connect the principle of liberty to the multiple cultures that exist in American society and consider its implications for the current standards-based approach to the reform of and accountability in public schools. Chapter five connects the principle of democracy in education to the politics of an overlapping consensus and considers how it helps to delineate a framework for civic education. Chapter six applies the principle of equal opportunity in education to the debates over common and differentiated curricula in schools and suggests how it can help to define an appropriate system of political control of these institutions. Chapter seven explains the consequences of the principle of economic growth in education for the economic education of children and develops an approach to funding schools to achieve the four principles of social justice in education. Finally, in chapter eight, I return to the controversy in Jamesville to consider how the entire conception developed and elaborated in the previous chapters can make sense of and resolve many if not all of the disputes in this hypothetical but typical American community.

The aim of this book is to develop a normative account of the politics of American education. This account will take seriously Americans' many declared aspirations for their public schools and reconcile those aspirations into a set of mutually consistent political principles for the conduct and outcomes of education. In turn, these principles will provide a critical and useful perspective on many current policies and practices in America's schools.

Chapter Two

A Political Theory of Social Justice for Education: Liberty and Democracy

In the previous chapter, we found that the citizens of Jamesville, the community's school professionals, and the local and state authorities disagree on a whole host of issues—ranging from the fairness of the joint federal and state system of school accountability, the sufficiency of the funding and other resources provided to support Jamesville's schools in responding to that system, and the adequacy of Carrington Middle School's response to the results of that system in light of recent changes in the school's demographics. We also saw that a theory of social justice based on John Rawls's idea of an overlapping political consensus might be a promising way to resolve at least some of those disagreements in a way that parties could accept, despite their many differences. The task of this and the next chapter is to ascertain whether an overlapping consensus about the principles of social justice in education is possible and, if so, to specify the content of those principles.[1]

John Rawls (1993) differentiated two ways of developing a theory of social justice—those that he labels political and those that he labels metaphysical. Metaphysical approaches to justice are based on comprehensive ethical theories, those that speak to the full range of human belief and experience and may include systematic religious or philosophical premises. Such approaches, I suggest, are implausible for a country such as the United States where citizens have not only commitments to different and sometimes conflicting fundamental beliefs but also a shared conviction that they have a right to hold and pursue different conceptions of the good based on the beliefs that seem reasonable to them. Political approaches, as already noted, emerge from an effort to identify an overlapping consensus about government among the normative beliefs of those who hold differing comprehensive ethical doctrines in a particular society. Using such an approach involves not only ascertaining the normative beliefs and judgments concerning government structures and operations about which a wide consensus actually exists. But it also involves submitting those consensual beliefs to a process of analysis in which the underlying conditions for the agreements are determined and, if necessary, good reasons are sought for modifying or restricting the beliefs to resolve, at least tentatively, any potential conflicts among them.

This approach depends on two basic ideas. First, people who hold different conceptions of the good may find themselves in agreement about political principles but for very different reasons that are grounded in their specific conceptions. Thus, for example, one group of people may think that personal freedom is important because they hold a philosophical belief that guaranteeing self-determination is a primary means to respecting the basic dignity of all human beings. Another group may think that personal freedom is important because of their commitment to worship their deity in the specific way their religion dictates and because they are concerned that political interference with their religious duties would cause them to violate a divine command. Obviously, these two groups have very different reasons for their beliefs in personal freedom, but they still can agree in general that such freedom is politically imperative. Second, however, because these ostensibly common political beliefs have such different rationales, they can lead to conflicts about the interpretation of the meaning of particular shared political values in concrete circumstances. Thus, those with the first conception of the good are likely to believe that personal freedom requires that parents grant their children autonomy in determining their own specific religious commitments, if any. But those with the second conception believe that personal freedom should protect parents' right to indoctrinate their children into the specific form of worship that they practice because otherwise they would be violating what many believe is a mandatory religious obligation. Despite their agreement on the value of personal freedom for adults, their different conceptions of the good lead them to disagree about the applicability of personal freedom to children. To some extent, we can see an analogous disagreement over school accountability in the situation at Jamesville, where both Latino and Anglo parents agree that parents should have significant influence over the education of their children, but where their influence is affected in different ways by the accountability system. We are not yet armed with the arguments that would enable us to resolve either of these conflicts, but it is clear that an analysis of the initial agreement to identify the underlying political value, if any, of personal freedom is necessary to finding such a resolution. Moreover, it should be clear that such an analysis could not embrace either of the contending parties' conceptions of the good, for politically acceptable resolutions must provide an interpretation of the meaning and extent of personal freedom and school accountability that all parties can reasonably accept.

One possible way to ascertain the overlapping political consensus in the United States about schools, if there is one, would be to survey all the various specific beliefs Americans have about education and then to submit all of those beliefs to a process of analysis that attempts to articulate the content of the manifold specific beliefs and the underlying commitments that can account for them.[2] However, given the large and changing variation of belief in this society, this strategy is not likely to be feasible. A more manageable approach is to determine whether any general purposes figure prominently and frequently in Americans' political discussions of their schools and then to ascertain whether those ostensibly shared purposes can become the basis of normative principles for public schools that constitute an overlapping consensus.

In this and the following chapter, I follow this second strategy. First, I note that four different purposes of schools have the requisite quality of being widely used in political justifications of school policies or changes in those policies: personal liberty, democracy, equality of opportunity, and economic growth. Then I analyze the idea behind each purpose and attempt to formulate a principle of justice to represent that idea in a way that applies to the operation of schools and that minimizes the conflicts with the other purposes. This approach, if successful, promises to generate a broadly appealing account of social justice in education and thus a basis for conducting a normative analysis of current school policies and practices that citizens of the United States can recognize to be consistent with how they think in general about their schools.

Before beginning the attempt to construct this normative political theory for schools, several remarks are in order. First, this will be a political theory specifically for public schooling; as such, it will not cover all of the territory about which an overlapping political consensus among Americans might be possible. It will not, on the one hand, consider political principles in the achievement of which public schooling seems to have at most an indirect role, such as those dealing with the distribution or redistribution of income and wealth. In addition, this theory will not, on the other hand, attempt to speak to all of the institutional arrangements necessary for achieving the political principles it does consider, such as the specific system of taxation to support schools. In short, only the public educational arrangements will be of immediate concern, and even then, it will treat only the most general political principles of public schooling rather than the full details of their implementation. In at least these ways, the resulting theory will be incomplete. Second, in constructing this theory, I will not develop what might be called an ultimate justification of the principles it includes, for such a task would involve attempting to ground this theory in comprehensive ethical premises, something that this method tries to avoid because of the inherently controversial character of such premises in a broadly pluralistic society. As a result, the theory—either as a whole or as any of its individual principles—will not necessarily be the most favored according to all of my own or, I suspect, anyone else's fundamental commitments. Third, I will, therefore, offer what attempts to be a political justification of this theory, one that appeals to the consensual nature of its principles. As noted, this way of proceeding assumes that people can agree on political principles even though they might disagree about the content and configuration of fundamental normative premises to which their agreement, in individual cases, is connected. Fourth, however, this procedure will not necessarily yield principles that formulate the commonly held political purposes of education in their most familiar form. For in that form, these purposes may well conflict with one another or be internally inconsistent, and it may be necessary from time to time to adjust the formulation of the principles based on them to ameliorate those problems. In doing so, I attempt to keep in mind the intuitive meaning of the various purposes and the relationship that seems to exist among them in the current political discourse about schooling. Moreover, I attempt to justify, or at least explain, any modifications in the formulation of the purposes based on the logical or practical necessity of such modifications to avoid self-contradictions in the larger theory that emerges. I also try to show that such modifications have desirable, or at least tolerable, implications for the society and its educational institutions. In the remaining sections of this chapter, I consider principles that might be derived from the first two widely discussed purposes of public schooling, namely, personal liberty and democracy.

Personal Liberty

I have already suggested to some extent just how personal liberty might be relevant to the situation in Jamesville. After all, Jamesville's parents, among other things, can be understood to claim that they should have the freedom to determine the content and procedures of their own children's education, and this is a claim that their personal liberty should include a right to control the purposes and the means of education, at least on the occasions when their children's interests are at stake. Of course, as we have noted, the interests of many others in the community, the state, and the nation are often affected by the schooling arrangements at Carrington Middle School, and therefore the personal liberties of many other citizens can be seen to be involved as well. One purpose of this section is to determine how to adjudicate competing claims to personal liberty when those claims conflict in the context of public schooling.

Liberty can be understood as the freedom to decide matters that affect our lives (Bull 1984 and 2000b). In the United States, citizens tend to treat liberty as a set of individual freedoms of self-determination, many of which are protected by the Constitution's Bill of Rights—freedoms of religious belief, association, and expression, for example. Thus, one aspect of liberty—personal liberty—implies one's control over a sphere of actions and decisions the primary effect of which can be understood as being on one's own life.

Of course, much, and perhaps most, of what we want to do has potentially significant effects on others' lives. In these cases, liberty is not the freedom to make these decisions entirely on our own but rather to participate fairly in making these decisions with others whom they also affect. Thus, liberty theoretically encompasses not only individual rights but also political rights, that is, rights to be involved in making the collective decisions that affect us. I consider political liberties only briefly in this section, but, because they are intimately involved with the political value of democracy, I reserve full treatment of them to the next section devoted to that topic.

At least four questions about the meaning we attach to liberty are important in developing a politically defensible conception of the personal liberties and in formulating a legitimate principle of the public schools' responsibilities relating to them—two that are of general import and two that are related specifically to the connection of these personal liberties with public schooling:

- What is the scope of the personal liberties that should be protected by the society because they are politically significant?
- What should happen when the exercise of one person's politically significant personal liberty affects the exercise of others' politically significant liberties?
- Is a political commitment to personal liberty consistent with a compulsory system of public schools?
- If so, how should such a system of schools be constrained by and promote a legitimate political commitment to personal liberty?

I consider these questions in this order because the answer to earlier questions is relevant to developing a response to subsequent questions.

An expansive conception of personal and political liberty leaves no room for the application of other political values. After all, every human action, no matter how momentous or trivial, can be understood as the result of someone's decision. If the right to make those decisions without interference is reserved to the individual or group primarily affected by them, it will be impossible to regulate any decision and therefore any action in the interests of achieving other political values. But the presence of equal opportunity and economic growth in the constellation of Americans' political values about schooling suggests that we do not hold such an expansive view of liberty. One's conception of politically significant liberty must, in other words, be constrained in some way to make room for other political values. The U.S. Constitution and its Bill of Rights provide some guidance on how Americans formulate this constraint. On the one hand, the Constitution provides for the protection of certain fundamental individual liberties—the freedoms of speech, religion, and association, for example. On the other hand, it provides a framework for what has been called the rule of law—for example, rights to the due process of law and against unwarranted searches and seizures.

John Rawls (1999b) suggests that what makes certain types of decisions and actions fundamentally important to individuals and, therefore, deserving of political protection is their centrality to individuals' personhood—that is, their playing a significant role in an individual's holding a conception of the good that he or she regards as personally meaningful and in having a reasonable chance to live a life governed by that conception. Thus, the freedoms of conscience and expression are politically significant in that they enable individuals to learn about the possibilities for, to formulate and revise, and ultimately to embrace conceptions of the good that have personal significance for themselves. Other imaginable freedoms, however, that do not have such an important effect on one's personhood—such as the freedom to drive on the left side of the road—are not fundamental in this way and will not necessarily receive political protection. Moreover, the protections of the rule of law, while allowing for the regulation of many individual decisions and actions that are not fundamental to personhood, are politically significant in that they ensure that the provisions of such regulation are publicly known in advance, explicitly formulated, and universally rather than capriciously enforced, so that individuals can, within known limits, confidently formulate and carry out life plans in accordance with their own conceptions of the good.

Now, even if the politically significant and therefore protected personal liberties are limited to those that have a central role in one's personhood, it is still possible for those liberties to come into conflict when multiple individuals attempt to exercise them. At least two cases of conflict are possible. First, one person's exercise of a protected liberty can diminish the chances that others have to exercise that same liberty. For example, one's exercise of the freedom of conscience might lead one to adopt an intolerant religious view that implies that one should restrict the religious views that others may adopt. In turn, acting on such a view would require one to limit others' freedom of conscience. Second, one person's exercising one sort of protected liberty can diminish others' chances to exercise a different protected liberty. For example, the exercise of one's freedom of conscience might lead one to adopt a sexist social doctrine that implies that women should not be allowed to fraternize with men. Acting on such a view would require one to limit some others' freedom of association. John Rawls (1999b) also provides guidance about the political resolution of such conflicts by suggesting that the aim should be to maintain a system of personal liberties that is equal for all. A system that protects the equal freedom of conscience or association, thus, may permit the restriction of even fundamental liberties in the interest of everyone's enjoying the same liberties. This arrangement does not necessarily mean that one is forbidden, for instance, from holding views that conflict with others' protected liberties (by authorizing, for example, the immediate incarceration or brainwashing of those who hold such views). But it does imply that one who holds such views may be prevented from or punished for acting on those views in a way that would effectively diminish others' protected liberties.

A compulsory school system might seem at first blush to restrict personal liberty inappropriately. On the one hand, a compulsory school system prevents children from pursuing the activities they prefer under their current, perhaps fragmentary and incomplete, conceptions of the good. On the other, such a school system may interfere with some parents' efforts to raise their children in accordance with the parents' conception of the good. However, it is improbable to maintain that young children, while they clearly have desires and preferences, are guided by a deliberatively chosen and consciously formulated conception of the good. Thus, restrictions placed on children's current actions do not necessarily violate their fundamental liberties because those actions are not necessarily expressions of their fundamental personhood. And the younger children are, the less likely it is that such restrictions will be forbidden by a commitment to their personal liberty. In fact, this circumstance is what renders parental authority over children acceptable in a society committed to the protection of fundamental personal liberties, a form of authority that would not be permitted among adults.

Parents, however, do have a protected liberty interest in raising their children; that is, they have conceptions of the good that usually have clear implications for how they exercise their authority over their children. Nevertheless, two considerations suggest that they may not exercise this authority in just any way that their conceptions of the good might imply. First, children are not simply parts of the environment that can be used willy-nilly as instruments for the fulfillment of their parents' conceptions of the good. Rather, they are nascent persons who can come to hold and pursue their own conceptions of the good, and thus they have liberty-related interests in the development of those conceptions (Bull 1990). Second, children also are members of the community whose aspirations and actions may affect the liberty interests of adults other than their parents—as, for example, future citizens or work associates. Thus, other adults also have a liberty interest in how children are raised. Constraints on parents' protected liberties in raising their children must be justified on the basis that such constraints are necessary to maintain an equal system of liberty for all—for the parents themselves, the adults whom children will become, and other adults outside the family (Bull, Fruehling, and Chattergy 1992). Thus, we have reason to limit the personal liberties of Jamesville's parents to educate their children as they choose even though such choices may be intimately connected to the fulfillment of the parents' conceptions of the good. However, these limitations are to be restricted to the extent that those choices interfere with other adults enjoying an equal liberty to pursue their conceptions and, especially, that they interfere with the future liberty of children to hold and pursue conceptions that the children as adults find personally meaningful.

A compulsory public school system governed by appropriate purposes and conducted in appropriate ways can be seen as one vehicle for maintaining such an equal system of politically significant personal liberties. For, first, such a system can foster young adults who are their own persons, in that they have personally meaningful conceptions of the good, are capable of pursuing them, and thus have the capacity to exercise their politically significant personal liberties. Second, the school system can foster young adults who are cooperative members of their families and the larger community, in that their conceptions of the good and their consequent activities are reasonably compatible with the scheme of activities dictated by the various conceptions of the good held by their parents and their other fellow citizens. It is unlikely, however, that a school system that takes exclusive direction from the liberty interests of either the parents or the other adults in the community at large will have such a result, for individual children's interests and emerging conceptions of the good must also be taken into account (compare Gutmann 1999). Third, the school system can also foster in children a responsible and thoughtful respect for the political values of their society, including personal liberty. Some of the details of the governance and operation of such a system will be the subject of chapter four, but based on this analysis, we are able to formulate one general principle of the normative political theory for education that we seek: *Conduct public schooling in a way that allows children to develop both as their own persons—that is, to come to hold personally meaningful conceptions of the good and to acquire the reasonable capacities to pursue them—and as responsible members of their families and communities who respect and support others' politically significant personal liberties and the other political commitments of their society.*

By following this principle, a compulsory public school system recognizes and respects the politically legitimate personal liberty interests of children, parents, and other adults simultaneously. In failing to follow it, a public school system cannot justify its compulsory nature because it may violate the current fundamental personal liberties of parents and other adults or because it fails to provide to children the prerequisites for exercising their future fundamental personal liberties as adults, or both.

Democracy

To a considerable extent, the situation in Jamesville involves a controversy over who should have political control over the community's schools, with some parents maintaining a right to control their children's education because of their personal liberty and others maintaining a right to such control in part because of schooling's potential effect on their personal liberty. We have concluded so far that personal liberty calls for a balancing of the interests of the parties to the extent that they affect the fundamental personhood of those involved in an effort to maintain the equality of the politically significant personal liberties. However, we have also seen that a good deal more is implicated in this situation than the personhood of the parties—that is, their rights to develop and pursue personal conceptions of the good that they find reasonable. Their interests in political liberty, equality of opportunity, and economic growth are involved as well. One purpose of this section is to consider just how extensive the claims to control education should be when they arise from citizens' membership in a democratic political order.

As I have suggested, democracy, at least in part, requires a recognition of individuals' political liberties, that is, their rights to participate in making the decisions that do not affect their fundamental personhood but that influence the course of their own and others' lives, decisions that extend beyond the purview of those involved in personal liberty. Here two questions seem important in determining the nature and extent of the political liberties that are politically significant and in articulating a principle according to which public schooling is to respect and promote those legitimate political liberties:

- Should all decisions that do not affect the politically significant personal liberties be subject to the participatory decision making process implied by political liberty?
- How should public schooling be arranged so as both to respect any politically legitimate limitations on political liberty and to foster politically significant political liberty?

The discussion of personal liberty and its significance to the personhood of individuals implies that political liberties must be restricted to prevent others from exercising an illegitimate influence on the conceptions of the good that individuals develop and embrace and on the capacities they develop to pursue those conceptions. After all, if others have an unlimited power to control the values we come to hold and the activities in which we engage, no one would ever be his or her own person, and the personal liberties would be meaningless, even if that power were exercised by means of participatory decision making procedures.

Of course, this does not mean that others cannot have any influence on our conceptions of the good and their pursuit. As we have seen, others can legitimately constrain such pursuit and even the conceptions themselves in the interest of maintaining an equal system of personal liberty. For these constraints allow two things to occur. First, they provide others the opportunity of informing us about possibilities for our conceptions of the good that they have found compelling. Second, they give us the chance to develop a conception of the good that acknowledges and respects others' conceptions and therefore to coordinate our activities with others in a way that allows the mutual if incomplete fulfillment of our conceptions. Despite these legitimate constraints, however, others cannot have the final say over individuals' specific conceptions of the good and their consequent life plans; that ultimate authority is reserved to individuals themselves.

Thus, the politically significant political liberties are limited by the politically significant personal liberties. Moreover, three considerations suggest that the legitimate political liberties are even more limited in scope. First, as already noted, Americans have allegiances to political values beyond personal and political liberty, namely, to equality of opportunity and economic growth. Thus, the range of decisions that fall within the scope of political liberty must be constrained to permit the realization of these other political values. I postpone consideration of these constraints to the sections of the next chapter devoted to those values.

Second, some fully participatory decisions made today can undermine political liberty in the future (Gutmann 1999). One obvious example is that a majority of citizens today might vote that some current citizens—perhaps, the poorest, those who do not speak English, or those who espouse particular political views—will not have the franchise tomorrow. Less obviously, citizens might adopt a policy that, while it technically does not exclude others from participation in political decisions, seriously reduces the likelihood that they will do so—for example, a poll tax set at a percentage of one's annual income or a requirement that all political debate is to be conducted only in English. Clearly, to maintain genuine democracy and therefore to respect political liberty equally, such decisions must be placed beyond the reach of democratic decision making.

A third more complicated but related consideration is that democratic decision procedures can be used to make decisions that undermine the culture necessary for the practice of democracy itself. Among such decisions might be to adopt what I have called authoritarian democracy (Bull 2000a and 2000b). Briefly, the authoritarian account of democracy holds that a popular government can and indeed should make all of its decisions in light of a robust and fully specified conception of the good for the entire society, a conception of the ends that the society seeks to achieve. In arriving at this conception, procedures of universal participation and thoughtful deliberation are to be followed. Of course, the social conception of the good may also be revised by following democratic procedures. In this sense, the decision making is democratic. However, once a conception of the good for the society has been adopted, all subsequent decisions about social policy and institutions are to be made in light of that conception; these subsequent decisions are interpreted simply as judgments about the most effective social means to the social ends specified in the conception. In this sense, this view is authoritarian in that the resulting conception of the social good becomes the sole authority for making subsequent decisions.

This view of democracy is appealing for several different reasons. First, it embodies a common understanding of instrumental rationality, namely, that one is rational by choosing coherent ends and then determining one's actions to accomplish those ends. Second, it forbids the use of the instruments of public authority for any purpose except for those articulated by the conception of social good that has been adopted democratically, so that no one will be permitted to use the power of the government exclusively for his or her private ends. Finally, it provides a straightforward basis for determining and judging public policy, namely, by telling us to develop the material, intellectual, and practical capabilities of the society in ways that will allow us to accomplish our collective purposes as reflected in the authoritative conception of the social good.

However, by focusing exclusively on the ends that the society is to achieve, authoritarian democracy seems to ignore political values that place restrictions on the means that may legitimately be taken in pursuit of our ends. One such value is liberty—both personal and political. A society committed to this value, as we have seen, cannot pursue its ends by violating the rights necessary to developing and pursuing the personhood of individuals or the universal participation of citizens. Another such value is equality of opportunity. As we see in the next chapter, a society committed to this value cannot pursue its ends by, for example, relegating some of its citizens to a permanently inferior social status based on characteristics that are irrelevant to their potential and achievements. This objection, however, simply requires us to constrain authoritarian democracy in appropriate ways but not necessarily to abandon it entirely. In ruling out the choice of means that entail a violation of protected liberties or of equality of opportunity, an authoritarian democracy may in some instances be forced to adopt means that are not the most efficient for attaining its ends. It may also turn out that some conceptions of the social good cannot be pursued at all because any means available to pursue them would violate these political values. Nevertheless, with these restrictions, a version of authoritarian democracy still seems attractive on the grounds already mentioned— its instrumental rationality, its forbidding public power to be used for exclusively private purposes, and its providing a clear criterion for the development of public policy.

For all its superficial appeal, however, democratic authoritarianism is a deeply flawed conception of political organization. In large part, these flaws stem from the empirical implausibility of its account of the connection between political will and social capacity. This theory essentially tells us to imagine the preferred future state of American society and then to engineer Americans' social capacities to achieve that state of affairs. However, this view does not recognize fully the extent to which our ability to imagine our future and, thus, our current choices of social ends are dependent upon our current social capacities. To be sure, it does recognize that our existing social capacities can serve as a reality check on our imaginations by telling us either that some visions of our future might be unattainable because of seemingly intractable limitations in our capacities or that some visions might have to be delayed while we develop capacities that we do not currently possess. However, it does not recognize two other important truths about the relationship between political will and social capacity: First, what we want or can imagine at any particular time depends on what we can do. For example, the people of Europe in the Middle Ages simply could not have held a conception of the social good that involved their being able to manufacture and deploy nuclear weapons. Second, what we will want in the future depends on changes that have taken place in what we can do. Thus, the relationship between political will and social capacity is interactive and not simply hierarchical, as democratic authoritarianism seems to assume. In other words, the supposition that we can come to a collective agreement about a vision of our desired social future, no matter how participatory and deliberative the procedures used to reach that agreement, and then simply change our capacities to achieve that vision must be false. After all, the subsequent changes we make in our capacities to achieve the original social vision will, in turn, change the social vision of the future that we deem to be desirable.

But perhaps this problem can be overcome by simply accepting that our current social aspirations are likely to change in the future. In other words, we might simply change our capacities today to achieve what we think we want tomorrow in the full recognition that what we will actually want tomorrow may be different and, thus, may require us to change our capacities in yet other ways. But even this adjustment in democratic authoritarianism is inadequate; indeed, it reveals an even more fundamental flaw in that conception of democracy. For this revised version of democratic authoritarianism is essentially a recipe for an unplanned, deeply irrational future. From a larger perspective, we can see such a society as profoundly schizophrenic, both believing that it should pursue a particular vision of its future and knowing that attaining that vision will inevitably prove unsatisfactory in entirely unpredictable ways.

In this way, decisions to pursue even a constrained form of authoritarian democracy, though made in a fully participatory manner, can make democracy seem frustratingly pointless, condemning the society to an endless cycle of choosing a social good only to have that good lose its apparent value when we have attained the capacities necessary to achieve it. In this sense, a choice to adopt such a scheme of political organization in the long run undermines citizens' very motivation to act democratically. To avoid this result, a democratic society must be constrained not just from making decisions that will undermine other significant political values and the future participation of some citizens in democratic decision making but also from implementing democracy in the centralized, instrumentally rational, but inevitably unsatisfactory manner envisioned under the authoritarian conception.

To this end, it would not be sensible to understand the overlapping consensus about democracy as the formulation of a unified and abstract national democratic will and its pursuit by engineering the nation's social capacity appropriately but instead as a set of more localized experiments in which a variety of concrete, competing, and incomplete hypotheses about the democratic will and social capacity are tested simultaneously. Some of these experiments may indeed take the form of attempts to attain particular goals by modifying social capacities to those ends to discover whether those goals still seem valuable to and attainable by citizens who have had their capacities thus modified. Others, however, may assert the putative value of a particular new configuration of social capacities to discover whether the goals that citizens come to have as they acquire those capacities seem worthy and attainable. Some of these experiments will succeed, and some will fail, but the communities that attempt them can learn from one another's experience about how to formulate more successful and satisfying subsequent experiments (compare Dewey 1927). And because of this experimental attitude toward the decisionmaking processes in their own society and the widespread availability of knowledge about the results of the variety of political experiments that continue to take place, citizens' motivation to be involved in democratic politics will remain robust.

It has been assumed, since at least the time of Thomas Jefferson, that democracy requires the operation of a public school system. In light of this analysis, we can see the connection between democracy and public schooling not merely as a way of giving future citizens the information and skills that they need to participate in the established decisionmaking processes and to carry out those decisions. We can also see it as an attempt to create and maintain a culture that accepts the various limitations on political liberty that I have argued to be necessary are implemented and respected. Understood in this way, a normative political theory implies an ambitious principle that should govern public education in society committed to the value of democracy: *Conduct public schooling in a way that fosters children's ability and willingness to participate in public decision making processes so that they acknowledge and respect the other political commitments of their society and so that they make constructive contributions to, learn from, and act on the results of those processes in both their own and others' communities.*

A public school system that follows this principle will encourage its students to exercise their political liberties in ways that respect the commitments to personal liberty, to their own and others' political liberty over time, to equality of opportunity, and to economic growth. At the same time, it will develop and maintain a democratic culture in which learning from one's own and others' political successes and mistakes is expected and valued.

Although this principle is not specific about the decision making arrangements appropriate to public schools, it and the other principles have clear implications for those arrangements that are explained in subsequent chapters, particularly chapters five and six. However, it should be clear that the central commitment of justice for education in a democracy concerns the way in which schools' decision making structures, instructional procedures, and curricular content enable and encourage students to prepare for and accept

their role in the operation and continuation of a suitably characterized democratic political order.

In many ways, the claims for political authority over schooling
made by the citizens of Jamesville, its state, and the nation neglect
this central aspect of social justice in education. Those citizens seem
instead more concerned about their own political power to determine the nature of children's education than they are about the
effects of such decisions on the political understanding and motivation that children come to achieve during and as a result of their
schooling. Of course, the application of democratic norms in the
making of decisions about the society's institutions is important,
but schools have implications for the future of democratic practices that few, if any, other social institutions have. As a result, it is
important to design the decision making arrangements for schools
carefully so that their consequences are compatible with the continuation and development of democratic norms in the society. Just
as the expression of parents' and others' personal liberties in schooling must be constrained to permit children an equal chance to have
and to exercise such liberties as adults, so too must parents' and
others' political liberties be constrained to maintain the conditions
necessary for children to exercise such liberties in accord with democratic norms when they assume the responsibilities of citizenship.
Thus, the simple fact that a political majority of Jamesville's adults
happen to prefer schools to teach in ways that are inconsistent with
all children's understanding and embracing democratic norms—or
with their enjoying personal liberties, equal opportunities, or a
legitimate role in the economy, for that matter—does not imply
that they have a right based on the exercise of their political liberties to enforce such schooling on the children of the community,
state, or nation. Beyond this, the operation of decision making procedures for schools also teaches children lessons about the purpose
and conduct of politics in their society. If, for example, the interests
of Latinos in Jamesville are allowed to be overridden by hardball
interest-group politics, no matter whether those interests align with

the society's commitments to fundamental political values, the community's children are likely to learn about the sincerity of the community's commitment to those values and about what is permissible in their own political conduct when they attain full rights to citizenship.

CHAPTER THREE

A POLITICAL THEORY OF SOCIAL JUSTICE FOR EDUCATION: EQUAL OPPORTUNITY AND ECONOMIC GROWTH

In a sense, considerations about the contribution of education to personal liberty and democracy speak to the quality of educational opportunity available within a society (Bull 2000a). Educational opportunity seen from these perspectives must be of the right kind to enable individuals to aspire to and to live lives that they find fulfilling and that contribute appropriately to the fulfillment of others' lives. Therefore, decisions about the kind of educational opportunities available in a society must be made considering their members' interests in personhood and citizenship. In Jamesville, these interests take the form of disagreements about the extent to which parents' and other citizens' aspirations and judgments should have an influence over the content and procedures of children's education. We have concluded thus far that schooling must be conducted in a way that enables children to develop as their own persons and as responsible citizens of a democratic political order with an appropriate political culture. In addition, parents' and other citizen's concerns about the content of the education provided by Carrington Middle School include the possible economic consequences of that education, consequences for individual children, the community, and the society as a whole. However, the claims of the parties to the disagreement over Jamesville's schools go beyond a concern with the content and control of children's learning opportunities. Latino parents, in particular, are concerned with the fair distribution of those opportunities as well in that they contend that the school system has previously neglected their children's achievement. To an extent, the reforms undertaken in light of the results of the state accountability system force Jamesville schools to pay greater attention to their children's performance in ways that they hope will correct for that past neglect. In this chapter, we complete the theory of social justice in education by considering how the distribution of educational opportunities and how economic issues should be best accommodated in that theory.

Equality of Opportunity

Concerns about equality of opportunity can be divided into those related to economic opportunity and those related to educational opportunity (Rawls 1999b). In this section, I focus on educational opportunities, although some consideration of economic opportunities will be unavoidable.

As is clear in the Jamesville case, in addition to caring about the nature of the opportunities that schooling makes available to children, Americans care about the fair allocation of those opportunities. If, for example, citizens decide that making a certain kind of educational opportunity available would be beneficial to them collectively, the chance to take advantage of that opportunity should be made fairly available to all. A fair chance to take advantage of an educational opportunity means in part that citizens should be allowed to qualify for that opportunity on the basis of their demonstrated abilities. But it also means that citizens should be given a fair chance to acquire the abilities that would allow them to qualify.

These two elements of equality of opportunity in education—a chance to acquire abilities and a chance to qualify for opportunities based on demonstrated ability—are the basis for a conceptual distinction in a society's educational institutions. In part these institutions are to be universal, giving all citizens a chance to develop their abilities, and in part they are competitive, giving all a chance to use their developed abilities to qualify for further education (Gutmann 1999). Very roughly, elementary and middle schools are the institutions that Americans expect to provide universal learning opportunities; high schools and colleges are to provide competitive learning opportunities.

It is natural to assume that public schools, in having a universal clientele and serving a universal mission in education, are to generate equal educational outcomes defined in some way, for such outcomes might be understood as a measure of the equality of the opportunities that children have experienced. However, two questions about this assumption are important to ask:

- Does such uniformity of learning outcomes when understood in the context of American social and economic institutions and Americans' other political values genuinely satisfy the requirements of equal educational opportunity?
- In light of the answer to this question, what does equality of educational opportunity require of this society's public school system, and how can schools reasonably attain it?

There are at least three possibilities for interpreting what equal educational outcomes could mean as an index of whether children have been afforded equal educational opportunity—first, that all children are to attain the same learning outcomes as a result of their schooling experience; second, that children with the same potential are to attain the same school outcomes; and, third, that all children are to achieve at least a minimum threshold of outcomes in certain socially and personally important subjects.

Amy Gutmann (1999) argues that the achievement of absolutely equal school outcomes is not only practically difficult to attain but also, and more importantly, normatively unacceptable. Although she bases her argument on a central commitment to the value of democracy, a reasonable approximation of it can be stated for the more eclectic political theory being developed here. She observes that differential student achievement derives, in large part, from the widely divergent educational resources to which families and communities have access and that they choose to devote to the learning of their children. Completely correcting for these differences would require massive collective intervention in the lives of these families and communities, which, as a practical matter is an extremely challenging task. Of course, perhaps a society should undertake this difficult task if persuasive normative arguments lead to that conclusion. Therefore, let us review the normative status of such intervention.

To the extent that these differences in student achievement depend on differences in their families' and communities' access to resources or differences in government allocation of educational resources to certain communities, they seem to represent a genuine injustice that should be corrected. After all, children are not responsible for choosing the families or communities into which they are born, and therefore they cannot be said to deserve the educational differences they experience, especially since those differences are likely to affect their developing personhood, that is, their developing conceptions of the good and their abilities to live their future lives according to them. But to the extent that these differences derive from choices that families and communities make on the basis of their individual and collective conceptions of the good, they are normatively more acceptable. For families' and communities' educational choices are also an expression of their members' conceptions of the good, which, within limits, are protected by their personal and political liberties. Thus, the educational differences that children currently experience can be seen to result, in part, from politically illegitimate violations of children's personal liberties and, in part, from the politically legitimate exercise of families' and communities' personal and political liberties.

As we have seen, the personal liberties of parents can, to an extent, be overridden in the interests of protecting the future personal liberties of their children. In other words, parents can be compelled to send their children to school to provide those children with a chance to become their own persons. However, even under a compulsory school system, parents still retain the right to contribute to the education of their children and, thus, to exercise, to a limited degree at least, their personal liberties in this regard. To produce absolute equality of student achievement, the society would have to deny altogether parents' rights to affect the education of their children, which would completely ignore parents' personal liberties in regard to their children's education. A similar argument about the political liberties of community members could also be constructed, which would conclude that full equalization of student achievement also denies communities the right to influence the education of their children. Thus, such intervention to produce completely equal outcomes, as the first interpretation of equal opportunities as equal outcomes requires, is not consistent with the political values we have developed thus far.

The second interpretation of equal outcomes holds that only children with the same potential should achieve the same outcomes in school and that those with different potential should achieve different outcomes. As a corollary, this interpretation often holds that, although potential varies widely among individuals, it is likely to be distributed equally among different groups of children, such as ethnic, racial, or gender groups, and therefore achievement should be the same across those various groups. As with the previous interpretation, the difficulties here are both practical and normative, and they become apparent when the precise meaning of children's potential is specified. One possibility for the meaning of potential is to see it only as native abilities. Not only is it difficult to find noncontroversial measures of all the diverse current talents that children may display (talents not only in standard academic subjects but also in music, athletics, and leadership, to choose a few examples), but it also is hard to distinguish when children's actual performances, on which judgments of their underlying potential must be based, are the products of their native ability or their social advantages or disadvantages. However, making such distinctions seems to be required on this interpretation of equal opportunity since opportunities are to be distributed in accordance with that underlying potential but not with children's social circumstances. If the definition of potential is expanded to include motivation as well as native ability, the problem of measurement is compounded since it may well be impossible to distinguish the portion of children's motivation that arises naturally from the portion that reflects the upbringing they have received in their families and communities. But even if these practical problems of definition and measurement were somehow solved, serious normative problems would still arise in the context of Americans' educational values that we have already discussed. Supposing these conceptual and measurement problems were solved, one can imagine that young children would submit to a battery of sophisticated tests of their potential talents and motivations and that then they would be allocated precisely the educational experiences that are appropriate to their measured potential. In such a scenario, it is hard to see just what role, if any, that children's emerging personhood and citizenship would play in their educational placements. In effect, the testing and placement system would determine their identities as persons as citizens for them. But for government authorities to make such a determination of children's identities would be to deny children's interests in their personal and political liberty. Thus, the second interpretation of equal educational opportunities as equal educational outcomes is also inconsistent with the political values we have developed.

On the basis of a similar analysis, Amy Gutmann (1999) suggests the third interpretation of equal opportunity as equal outcomes, namely, that a more realistic and morally justified interpretation of equal educational opportunity, rather than aiming for absolute equality of student achievement or equality of achievement for those who have similar potential, requires us collectively to ensure that every educable child meets a minimum threshold of school achievement in the subjects that are universally instrumental to their success, whatever their own conceptions of the good may happen to be.[1] Such a threshold would recognize the interests of children in their future liberties by guaranteeing them a platform from which to make judgments about and to pursue their own futures. But beyond the threshold, parents and the members of communities would have the right to offer (or not to offer) additional educational opportunities to children, which would to an extent also recognize parents' and community members' personal and political rights.

I find this suggestion to be attractive, based as it is on the realities of American social life and on Americans' commitments to personal and political liberties. It, moreover, seems relevant to the current standards-based strategy of school reform and the accountability system that is of particular concern in the Jamesville case. However, I believe that two considerations suggest that this interpretation does not follow the implications of these realities and commitments to their full logical conclusions.

Let us examine, first, the ramifications of the pursuit of equal school outcomes up to a specified threshold, which I call equal basic education, for the lives of children while they attend school and as adults thereafter in a society such as that in the United States. Some children will have more difficulty in reaching this threshold than others. In part, this implies that the society will have to devote more and better focused resources to the basic education of these children than will be necessary for others. But it also means that these children will have to devote more of their time and energy to attaining a basic education than will other children. Moreover, based on current sociological realities, the children who have difficulty in meeting the threshold will come disproportionately from poor and socially marginalized communities and families. Unless we assume that all children will be required to leave the public school system when they meet the threshold, and Gutmann argues against this assumption, some children, by contrast, will have the time, energy, and public resources to pursue their education beyond the threshold. On top of that, these children will also tend to have disproportionate access to the additional resources that their parents and communities can afford to devote to their children's education beyond the threshold.

American children live in a competitive society and economy in which their chances to succeed as adults depend significantly on any comparative advantage they may come to have over their peers. Achieving up to the threshold will not confer a competitive advantage on anyone since by hypothesis all children are to reach this level. Thus, any educationally relevant comparative advantage that children attain will depend entirely on their achievement beyond the threshold. However, children from marginalized backgrounds will be further disadvantaged in their attainment of a comparative advantage for two reasons—they will be required to spend more of their time and energy in attaining the threshold, which as we have said confers no comparative advantage, and they will have less time and fewer resources—both public and private—for their education beyond the threshold, which does confer such an advantage. Ironically, then, establishing a threshold of school achievement that all children are to meet puts poor children at an even greater comparative disadvantage than they would face without the threshold, and the higher the threshold is set, the greater the social disadvantage of those who have difficulty in meeting it compared to those who do not (Bull 1996). To be sure, these children will be ostensibly assured that they can achieve at the threshold level in that their society will provide the resources to make such achievement possible, but achieving at that level does not enhance their chances of success in a competitive social and economic system. Of course, on Gutmann's interpretation of equal opportunity, there is no normative reason to be disturbed by this result since all children have been provided with an educational start in life that takes them at least to the threshold level.

That leads us to the second consideration—whether we should be satisfied with the threshold interpretation, especially given the formulation of Americans' political commitments that we have developed. The reasoning for the satisfactoriness of this interpretation seems to rest on one of two assumptions—either that the normative adequacy of an education is to be judged entirely on its own merits, not for any effect that education may have on children's subsequent social or economic success, or that, although an education's adequacy is to be judged at least partly on its subsequent social effects, an education up to the threshold can provide children with a reasonable platform for pursuing school achievements that will enable them to differentiate themselves from their peers in the competition for success. While the first assumption has a certain academic appeal in that it portrays education as having inherent rather than instrumental value, it goes too far in this direction by suggesting that education's political value ought to rest entirely on its inherent value. Moreover, in the context of the political values that we have formulated, judgments about inherent value are to be made from the standpoint of individuals or communities—based on their own personal or collective conceptions of the good—rather than by the political structures of the society at large. To the extent that education or any other good is a legitimate *political* concern of the society, it must therefore be understood as instrumental to the satisfaction of individuals' or communities' self-determined conceptions of the good, not as realizing any inherent good that the society deems to be authoritative. It is not that education must be viewed as bereft of inherent value, but the society must not be the judge of such value.

Furthermore, the logic of equal opportunity recognizes this distinction between inherent and instrumental values. As noted at the beginning of this section, equal opportunity has economic and educational aspects, but they are intimately related in that educational opportunity is of concern to the extent that it develops the realized abilities according to which economic opportunity is to be allocated. Conversely, not every educational difference need be of concern to the society, only those that lead to an unfair distribution of economic opportunities and social benefits. Only in this way, indeed, can a society allow for the differential education that may be required for individuals or communities to develop and to pursue different conceptions of the good, that is, for them to exercise their personal and political liberties. Therefore, the first assumption on which the argument for the normative satisfactoriness of the threshold interpretation of equal educational opportunity may rest—that the threshold for education is to be judged only for its inherent value—simply cannot hold for a society that is committed to the political values we have been formulating.

Perhaps the second assumption will prove more robust. It asserts, on the one hand, that there is a common set of learnings that everyone must achieve to have a fair chance of developing the comparative advantages that will lead to social success and, on the other, that the threshold guarantees everyone a reasonable chance to attain these common learnings. Of course, this is partly an empirical proposition in that it asserts, first, that there are certain common skills and knowledge on which everyone's social success depends and, second, that we know what they are. I find the first assertion to be wildly improbable for the sort of open and diverse society that is consistent with the political values we have described. Even what are taken to be basic literacy and computation skills seem not to be necessary for social success, as the many stories of real but improbable success in American society attest. However, the second assertion— that we know what these universally instrumental skills are—must surely be false. The course of scientific, technological, and economic development is an inherently unpredictable affair. As such, the claim that we know the prerequisites for such development with any more precision than, perhaps, the most truistic admonitions might lead—don't cause an atomic holocaust, for example—must be delusory. Beyond the improbability of the empirical content of this assumption, its normative content also seems inadequate. For it asserts that the society's educational obligations to its citizens have been satisfied when it has provided the bare prerequisites for success. Beyond that, any educational provision must be entirely a matter of individual luck—say, whether the family or community we happen to be born into values education beyond the threshold. Of course, educationally disadvantaged children may be so in part because their parents and communities, whatever their level of income, attach insufficient value to education beyond the threshold in general or to the particular sort of education beyond the threshold that will help realize their children's emerging conceptions of the good, and equality of educational opportunity is supposed to correct for such circumstances. However, it seems that the interpretation of equal opportunity as equal achievement up to a threshold cannot achieve these corrections. On both empirical and normative grounds, therefore, the second assumption on which the defense of the threshold interpretation of equal opportunity may rest seems as dubious as the first.

Of course, some other normative defense of this interpretation may exist, but it is not apparent to me at least. Therefore, I am driven to the conclusion that the third interpretation is normatively unsatisfactory in light of the conditions that obtain in a society committed to the political values that we have formulated.

This result could lead us to the uncomfortable conclusion that equality of educational opportunity may have no legitimate place in a society committed to the values of personal and political liberty. After all, neither the absolute equality of learning outcomes, the equality of such outcomes for children with similar potential, nor the equality of outcomes up to a threshold level is consistent with those values. However, it may be that all these interpretations are seeking equality in the wrong place—that is, in terms of equal learning outcomes. To see how equality of opportunity might be attained without uniformity of learning outcomes, let us consider the following argument that tries to establish a connection between the two.

Intuitively, the general idea of equal opportunity is concerned, among other things, with the future consequences of education—its effect on the chances at success that children will have in achieving, for instance, political power, economic well-being, and social status when they reach adulthood. Moreover, equal opportunity implies that those chances at social success are to be equalized in some meaningful but unspecified way. However, such equalization does not logically entail the uniformity of school learning. To reach this result, additional premises are required. One such premise might be that American society is characterized by a single ladder of success—in other words, Americans hold a common definition of personal fulfillment, and there is a single pathway to attain it. Coupled with this initial assumption, a second premise will complete the argument that equal opportunity requires educational uniformity—namely, that for climbing that ladder a person needs certain educational prerequisites. To equalize the chances at success, the society apparently will have to provide all its citizens with those prerequisites. For, although attaining those prerequisites does not guarantee success, such success cannot be achieved without them.

Liberty—personal and political—tends, however, to undermine the first premise (Nozick 1974). Within limits, personal liberty provides individuals with the chance to adopt, revise, and pursue their own personal conceptions of the good in terms of which they make decisions about the course of their individual lives. Again within limits, political liberty also provides citizens with the chance to develop and to change their collective conceptions of the good in terms of which they make their political decisions. These liberties, in other words, allow individuals and groups to develop and modify their own definitions of personal and social success. Against such a background, it is highly implausible that a single definition of success will emerge naturally. Of course, many individuals and groups may come to share common elements of their conceptions of the good, perhaps even common conceptions of the good in their entirety, but the membership in these agreements is constantly changing, and their precise content tends to evolve over time. Moreover, a society committed to these liberties is obligated to prevent individuals from using their social and political power to enforce a single authoritative definition of success on everyone else. Thus, on a combination of normative and empirical grounds, one element of the first premise in the above argument for uniformity of school outcomes is false, at least for societies that value personal and political liberty. Although I will not pursue them fully here, I find equally implausible several other elements of this argument. For example, even though individuals share a conception of the good, there may be many different ways for them to succeed according to that conception. Be that as it may, without the assumption of a single pathway to success, this argument for educational uniformity cannot succeed.

The failure of this argument, however, opens up at least one possibility that a normatively satisfactory form of equalization can be achieved without requiring uniformity in the school learning expected of and made available to children. This possibility needs to be explored as we search for a way of integrating equal opportunity with Americans' commitments to personal and political liberty. Rather than equalizing the opportunities that all children have to attain the same educational outcomes, we might equalize the opportunities that all children have to become different in ways that advantage them in the various competitions for further educational and employment opportunities that are meaningful to them (Bull 2000b). As long as the determinants of these educational differences come from children's emerging personal and collective conceptions of the good, a society that pursues this sort of equality will not diminish children's future liberties. And as long as the differences that we permit and support respect others' rights to develop and pursue their conceptions of the good, we also will not violate the existing liberties of adults. Since the principles governing education for personal liberty and democracy already limit the conceptions of the good that children are encouraged to develop to those that are consistent with others' personal and political liberties, the principle of equality of opportunity in schooling becomes: *Conduct public schooling so that children have an equal chance to develop the differential abilities required for success under their emerging individual and collective conceptions of the good.*

A public school system that acts on this principle will, to put it simplistically, give those children who aspire, for example, to become poets as good a start in that direction as it gives to those aspiring to be electrical engineers. This principle requires a society to exert something like equal educational effort on behalf of the self-defined success of each individual child as that society endeavors to meet the requirements of the first two principles. Indeed, as Foley, Levinson, and Hurtwig (2001) explain, this is what various scholars of color have been emphasizing recently as they identify and document the comparative advantages that children from minority communities bring to their education. Moreover, public schooling that supports a wide range of children's aspirations is more likely to lead to a more broadly based form of social equality than a narrow focus on an authoritative threshold of skills and knowledge, for, although different children will be prepared to do different things in their adult lives, they will be supported in developing the talents of greatest meaning to them. In a very real sense, a society's enforcing a single ladder of success and the public schools' complicity in doing so makes social inequality incorrigible both by narrowing the talents that children can develop and utilize in their search for individually and socially fulfilling lives and by allowing those who have current advantages to focus their resources on maintaining that advantage (Bull 2000a). Of course, the details of how schools can support the development of socially constructive differences among children are extremely important, and those matters will be addressed to an extent in chapter six.

On this view of equal educational opportunity, the Latino parents in Jamesville are expressing a politically legitimate concern about the past unfairness of how those schools have distributed opportunities to their children. But in assessing that unfairness, we must evaluate school policies and practices for the extent that they have enabled schools to make equal efforts to support the aspirations that children come to have as a result of their developing personal and collective conceptions of the good as implied by the first two principles. An exclusive focus on whether children have opportunities to develop and to achieve the aspirations that their parents, the other members of the Jamesville community, or the citizens of the state have for them is simply insufficient to meet the requirements of the third principle. Of course, to the extent that the aspirations of the young align with those of older generations, a failure to support the aspirations of adults for children may indirectly reflect a failure to achieve social justice. However, the critical element in assessing equal opportunity in education is whether the schools take with equal seriousness all the diverse aspirations of their students.

Economic Growth

Among the concerns of the citizens of Jamesville and its state are the economic consequences of students' school performance. Parents and other community members do care that their children's education enhances their earning potential, but that issue may not be among their immediate concerns. The real pressure for children's economic productivity is likely to be expressed by those at a greater remove from children, namely, those in the state legislature and the Congress who have enacted the standards-based accountability system. To some extent, the motivation at those levels may stem from the conviction that any educated person should be able to demonstrate the language and numerical skills included in the state standards. But that *these* subjects are the focus of school accountability, rather than music or citizenship, which are equally plausible candidates for subjects of which an educated person must have at least a basic mastery, makes it likely that much of the concern behind the accountability system lies with children's ability to contribute to the nation's future economic growth.

The most traditional and straightforward justification of the political value of economic growth is based on utilitarian ethical theory, the idea that a society should seek to maximize its members' happiness (Rawls 1999b). Although there are important controversies over the most adequate formulation of utilitarian theory—for example, whether the total or the average happiness is to be maximized or whether happiness consists in the satisfaction of desire or the realization of inherent value—the utilitarian argument for the value of economic growth runs something like this: Because humans are at least in part material beings, material objects play an important role in the attainment of human happiness by, for instance, directly satisfying human desires or realizing inherent value or by establishing the conditions under which the satisfaction of desire or the realization of inherent values can be pursued. Thus, the attainment of happiness depends significantly although by no means entirely on the material well-being of a society's members, that is, the material resources they have at their disposal. In general, the greater the supply of material resources that is under human control, the greater the chances for happiness. Because economic growth expands this supply of material resources, it is valuable as an instrument for maximizing human happiness.

Of course, there are significant variations among utilitarians about such matters as the correct distribution of economic resources and the social conditions under which those resources can best be developed, but one conclusion about this argument for economic growth is unavoidable—namely, that it cannot be used in its original form as a *political* justification for economic growth in the sense intended here, that is, as an appeal to a consensus that can be maintained regardless of one's fundamental commitments about what is valuable in life. After all, utilitarianism's assertion that human happiness, however it is interpreted, is the highest human value is itself such a fundamental commitment, and a highly controversial one at that. For Americans can and do disagree about whether or the extent to which happiness is the fundamental value in life, committing themselves to a wide variety of alternative accounts of ultimate value, such as obedience to the will of God, the development of human reason, the maintenance of caring relationships, and many more. For this reason, such a justification is inconsistent with the accounts that we have given of personal and political liberties and of equality of opportunity.

To be sure, various utilitarian theorists have argued that liberty, equality of opportunity, or both are instrumental to the achievement of human happiness. For example, John Stuart Mill (1859/1978) argued that since individuals are the best judges of what makes them happy, a society that aims to maximize happiness must allow individuals to make their own choices in life and to pursue the courses of action to which those choices lead. Similarly, a utilitarian might argue that equality of opportunity improves social efficiency by allocating citizens to the social positions for which they are most qualified and productive, and such efficiency permits the society to produce the most happiness with its available stock of resources. However, these utilitarian arguments are not entirely consistent with our formulation of the values of liberty and equal opportunity. For instance, the principle of personal liberty we have developed implies that individual commitments to fundamental values are largely matters of personal conscience, but the utilitarian arguments for liberty implies that a society must be committed to assigning the highest and most fundamental value to human happiness regardless of what its citizens happen to think. Similarly, the principle of equal educational opportunity we have developed specifies that the various educational opportunities that a society is to make available are to be guided by children's own emerging conceptions of the good, but the utilitarian arguments for equality of opportunity do not respect such a specification because they assume that social and economic opportunities are to be allocated so as to maximize human happiness regardless of whether the individuals and groups affected by that allocation happen to aspire to that end.

In light of these conclusions, the answers to two questions are critical to the formulation of a political principle for public schooling in the service of economic growth:

- What are the significance and value of economic growth, if any, in a normative political theory of education?
- What role do the public schools have in attaining the politically significant value of economic growth?

The principles of personal and political liberty imply a concern with both the quality and quantity of opportunity that is available in a society. Being our own persons requires, in part, having adequate resources to make judgments about or to develop our own conceptions of the good, but it also requires the resources to develop plans to achieve those conceptions and to carry them out to the extent that they do not unduly interfere with others having the resources to do so as well. To some extent, the resources required for personal and political liberty are protections of our freedoms of conscience, expression, and association and guarantees that the society will make decisions democratically and be bound by the rule of law in making and implementing those decisions. These resources, as we have seen, are also educational in that individuals require the chance to learn about the personal and collective conceptions of the good that they may find fulfilling, to gain the skills necessary to make good decisions about the adequacy of those conceptions, and to develop the abilities to construct plans to realize those conceptions and to carry them out. But beyond this, we also need appropriate opportunities to enact those plans. In other words, the opportunities available in American society should ideally be sufficiently numerous and of the right kind to give its citizens a reasonable chance to live their lives as their conceptions of the good dictate.

In any actual society, of course, these opportunities are inevitably inadequate to allow all citizens the luxury of realizing their conceptions completely. Thus, we will have to negotiate and compete with one another over who will have access to these scarce opportunities. We want this negotiation and competition to be fair, and it is the purpose of the principle of equal opportunity to ensure that they are by, on the one hand, giving citizens a chance to develop abilities that are in line with their conceptions of the good and, on the other, guaranteeing that the decisions about the allocation of opportunities to use those developed abilities are made according to citizens' potential to take advantage of them.

A society's economy is relevant to these concerns in two ways. First, it supplies the resources necessary to enforce citizens' liberties and to provide the education required by the exercise of those liberties and the pursuit of the ways of life toward which those liberties lead. Second, the economy generates many of the opportunities that citizens have to realize their conceptions of the good either by supplying the material prerequisites of the pursuits those conceptions imply or by creating types of employment that themselves realize their citizens' aspirations. Thus, for instance, for someone whose conception of the good requires climbing mountains, such employment may allow the accumulation of the resources needed to accomplish that feat. And, for someone whose conception of the good requires conducting scientific experiments, employment may be available in that very line of work.

Growth of the economy is an expansion of the productive capacity of a society. First, such growth can generate the resources needed for a society to provide more adequately for the protection of liberty and the delivery of the education needed to realize those liberties and to equalize opportunities. Second, economic growth can also enhance citizens' concrete opportunities to realize their conceptions of the good (Bull 1996). An expansion of the economy takes place in one of two ways, either by increasing the number of available jobs or by enhancing the productivity of existing jobs. When the number of jobs increases, there is more opportunity available for work in the society. When the productivity of jobs increases, the quality of opportunity in the society may be improved. In either case, the quality of life for the members of the society may be enhanced, not only because there is more disposable income available to them for the realization of their conceptions of the good, but also because the opportunity to engage in personally rewarding employment is expanded either to include more people or to enhance the work experience of those already employed. In other words, economic growth can be socially valuable because it transforms the quantity and quality of opportunity for earning income and for satisfying work in a society.

This argument for the social value of economic growth definitely is not an unqualified endorsement of that phenomenon, as the utilitarian argument tends to be. For a society's economy may expand without increasing the resources available for the enforcement of liberty or the provision of education. Indeed, some mechanisms of growth may actually diminish those resources either by making citizens less willing to pay taxes to support these things or by, for example, requiring child labor that displaces education. Moreover, either of the mechanisms described for the enhancement of productivity may not have the desired effect on citizens' ability to realize their conceptions of the good. That is, the jobs added to the economy may displace some higher paying jobs, or changes in the quality of jobs may make them less likely to satisfy the requirements of those conceptions. When, however, economic growth results in more public resources for the protection of liberties, the enhancement of education, or the equalization of opportunity and when it produces higher paying jobs or more fulfilling work, it is of genuine value within the normative political theory we have been developing.

Any legitimate principle of public education for economic growth must be framed within this account of the political value of economic growth. Of course, it may well be that beyond education for liberty, democracy, and equal opportunity, public schools have no additional role in enhancing politically desirable economic growth as we have defined it. These forms of education already make important contributions to the realization of this type of economic growth. Education for liberty enables children to become their own persons, which in turn makes it less likely that they will accept forms of employment that do not help them advance their conceptions of the good. Education for democracy enables children to understand their obligations within the society, which makes it less likely that they will use their political liberty to authorize forms of employment that obstruct their fellow citizens' chances to realize their conceptions of the good. Education for equal opportunity develops children's abilities in line with their emerging conceptions of the good, which helps to ensure that the paths most readily available for economic growth in their society are consistent with those conceptions. Despite these contributions, a society may neglect the opportunities available to encourage legitimate forms of economic growth. Moreover, it still seems possible for economic growth to take forms that are antithetical to Americans' most central political values, and society may legitimately act to protect itself from this possibility. The crucial issue is whether schools, beyond meeting the responsibilities we have identified, can be reasonably expected to be involved further in the actions necessary to encourage constructive economic growth or to discourage economic growth that is destructive to Americans' basic political commitments.

In the past several decades, economists have developed what has come to be called human capital theory (e.g., Thurow 1970). This theory notes that human capabilities are as important as factors of production as are raw materials and labor power, for example. Thus, improving economic productivity may require changes in the economic capabilities of citizens as much as it requires, say, enhanced access to natural resources. In fact, such capabilities can lead to improvements in our access to and use of the other factors of production. Human ingenuity can develop novel products or streamline production processes and resource extraction in ways that render the economy more productive. Indeed, this line of thinking lies behind many of the efforts to reform the public schools in America since the 1980s. For schooling has become thought of as an important social means of changing future workers' capabilities and thus of encouraging economic growth. On this theory, schools have a definite economic role in developing their students' motivations and abilities in line with the present and future needs of the economy.

However, our analysis of the legitimate political value of economic growth implies that Americans should be cautious about adopting human capital theory in its entirety, in large part because the normative justification of contemporary economics appeals straightforwardly to utilitarian ethical theory, which we have found to be at least partially inconsistent with Americans' other political values. For example, human capital theory might direct public schools to develop certain work skills among students even though their emerging conceptions of the good might lead them in quite different directions. Even worse, this theory might authorize schools to engineer students' personal and social conceptions of the good in particular ways that are deemed necessary to encourage economic growth. Finally, this theory might, on the grounds of efficiency, direct schools to neglect or minimize the education of certain segments of the student population—for example, students with disabilities—and to emphasize the education of other segments. Of course, these are only theoretical possibilities, but their very existence demonstrates that the normative grounding of the theory of human capital is at odds with the normative political theory that we have been developing here.

Nevertheless, when appropriately qualified, human capital theory can help clarify the economic role that schools are to play in the lives of their students. The schools' commitments to education for personal liberty, democracy, and equal opportunity are their highest priorities. But, as these responsibilities are being accomplished, schools have an obligation to inform students about the economic value of various developed capabilities, to help them create life plans that consider the role that such capabilities may have in pursuing their own emerging personal and social conceptions of the good, and to provide opportunities for developing any economically valuable capabilities that their life plans may include. In other words, schools should encourage politically legitimate economic growth by helping their students understand whether and how such growth contributes to the realization of their personal and collective conceptions of the good. Thus, a principle emerges for the public schools' politically legitimate role in enhancing a society's economic growth: *Conduct public schooling in a way that allows children to understand the role that economically valued capabilities may have in formulating and pursuing their emerging personal and social conceptions of the good and that helps them develop the economic capabilities included in their life plans.*

This principle allows public schools to contribute to their society's economic growth by honestly representing to students the personal and social economic consequences of their decisions and by fostering the economically valued capabilities that are of meaning in their students' lives. But it does not permit the schools or their political leadership to become the ultimate decision makers about which paths toward economic growth, if any, students are to follow. That right is retained by the students themselves. Of course, in following this principle, a society's economic growth is likely to be less than maximal, but it will be constructive growth in the sense of contributing to the achievement of its citizens' conceptions of the good.

In Jamesville, this principle reminds us to be cautious about the influence that the economic motivations of adults may have on the content, distribution, and instructional procedures of children's education. For the economic considerations that are of greatest normative importance are those relevant to the creation of the resources that are necessary to protect liberty and equal opportunity and, then, to realize the economic motivations of children themselves as they pursue the conceptions of the good that those liberties and opportunities make available.

A Conception of an Educationally Just Society

To recapitulate, an overlapping consensus about American education, I suggest, encompasses four political principles:

Personal Liberty

Conduct public schooling in a way that allows children to develop both as their own persons—that is, to come to hold personally meaningful conceptions of the good and to acquire the reasonable capacities to pursue them—and as responsible members of their families and communities who respect and support others' politically significant personal liberties and the other political commitments of their society.

Democracy

Conduct public schooling in a way that fosters children's ability and willingness to participate in public decision-making processes so that they to acknowledge and respect the other political commitments of their society and so that they make constructive contributions to, learn from, and act on the results of those processes in both their own and others' communities.

Equality of Opportunity

Conduct public schooling so that children have an equal chance to develop the differential abilities required for success under their emerging individual and collective conceptions of the good.

Economic Growth

Conduct public schooling in a way that allows children to understand the role that economically valued capabilities may have in formulating and pursuing their emerging personal and social conceptions of the good and that helps them develop the economic capabilities included in their life plans.

In the remainder of this book, I apply these principles to a variety of policy controversies in education. Before that, however, several brief observations are in order. First and most obvious, these principles do not make a definite commitment about the values that make life worth living, for those commitments may be various, and to the greatest possible extent they are left ultimately for the individuals who make up the society and for the communities in which they are involved to determine for themselves. In short, neither the conception of social justice in education nor the institutions that implement that conception, especially public schools, decide these fundamental issues of value. In this sense, the institutions of the society are neutral to the conceptions of value that their citizens hold.

Second, however, these principles do resist individuals' and communities' fulfilling certain conceptions of those fundamental values, namely, those that would deny the principles' commitments to universal personhood, universal political participation, equality of opportunity, and economic understanding among the society's citizens. Therefore, while the citizens of this society may choose to adopt discriminatory conceptions of the good that conflict with these commitments, the institutions of that society may forbid them to act or to use government power to achieve those conceptions if in doing so they make it impossible for the society to realize the four principles of social justice in education. Thus, for example, it may be allowable that some citizens hold a religious doctrine that deems others who do not accept that doctrine to be infidels and thus unworthy of citizenship, but the government may prevent them from teaching such a doctrine in the public schools. In other words, although government is to be neutral to the fundamental beliefs of its citizens, it is not necessarily neutral to all the actions taken on the basis of those beliefs. In this way, a socially just school may frustrate the values of some citizens even if it thereby discourages the adoption of certain conceptions of the good.

Third, although each principle is stated independently, the satisfaction of any single principle does not necessarily render the policy or practice of schools just; rather, social justice results only from the four principles operating together. Thus, a school that equalizes its students' opportunity but does not permit them to become their own persons or one that encourages students to develop their own conceptions of the good but denies them knowledge of the economic consequences of their options is patently unjust. To a great extent, this requirement is built into the statement of each principle in that each directly or indirectly refers to all the others. Indeed, the potential conflict between the principles and the need to coordinate and restrict them to prevent such conflict was a central consideration in the discussion of their precise formulation.

Fourth, these principles are not exclusively or even primarily individualistic in their orientation. For instance, although personhood is a central commitment, citizenship is of equal importance. Moreover, the source of one's personal and social conceptions of the good, and thus of one's personhood, is rarely the individual alone. Rather, many others, including a child's family, community, future employers, future workmates, and future political associates, contribute to the possibilities for those conceptions. What these principles forbid, however, is that these others make the final and exclusive determination of that conception.

Finally, decision making in a society governed by these principles is in many ways but not entirely decentralized. Certainly personal conceptions of the good are determined within reasonable limits by the many persons who hold them. Moreover, the second principle requires that most public decisions about the social good are to be made in the communities that are most affected by them. What may not vary among people or their communities, however, is the substance of the commitment to social justice in education as defined by the four principles together. Thus, while communities may, for example, decide differently how best to achieve equality of opportunity in their children's schools, they do not have the prerogative of not achieving it at all. Central decision-making authority, therefore, is to be exercised to ensure that these fundamental commitments are respected and enforced.

Chapter Four

Personal Liberty and Education: Families, Cultures, and Standards

According to the principles derived in the last two chapters, both children and the larger community have normative liberty interests in children's schooling. For children themselves, the education they receive should provide a reasonable basis for their emerging personhood. It must allow them to develop and pursue a personally meaningful conception of the good in such a way that they themselves are appropriately implicated in its eventual constitution. For the larger community, the education of children should enable and encourage them to acknowledge and respect the political commitments of their society—its commitments not only to others' personal liberty but also to democracy, equal opportunity, and economic growth.

Culture and Liberty

Nothing in these principles suggests that children must invent from scratch their own conceptions of the good for them to be their own persons. Indeed, it is highly unlikely that such individual inventions will be as potentially rich and meaningful as the conceptions developed over millennia by the world's varied and sophisticated cultures and civilizations. Thus, to become their own persons children inevitably must have access to such cultures, and, to meet the requirements of the principle of personal liberty, schools must consider the extent to which and the ways in which they have a role in providing that access.

Of course, the concept of culture is elusive, Protean, and widely contested, especially in academic circles. It is not my intention to stake out a position on these scholarly disputes but simply to rely upon an intuitive and nontechnical understanding of that phenomenon as it figures in the lives of citizens of the United States. Most often, culture is largely invisible to us. To a great extent, our cultures are simply life orientations that are revealed in what we think, how we act, and what we say as we are absorbed in our daily activities. Our cultures encompass such things as our languages, our conventional social practices, our usual ways of relating to and interpreting our natural and social environments, and our typical attitudes toward them. In many cases, our cultures include certain explicit and conscious values, such as those involved in our religious and political beliefs, but all these practices, relationships, and attitudes can be understood to comprise and reflect a set of deep normative commitments that are for the most part implicit, sometimes inconsistent with one another, and incomplete but that make subtle contributions to our emerging and evolving conceptions of the good. Moreover, we all come to have a culture in this sense even though we are not continuously and consciously aware of it or of how and when we acquired it. By simply growing up in a family and in a community, we ineluctably become acculturated. In many ways, culture is like the water in which fish swim—imperceptible but crucially important in our lives. Sometimes, however, when we encounter others who speak, think, and act in ways that we normally would not, we come to see them as having a culture, which on reflection means that we must have one, too. Such experiences are much more frequent and conspicuous for those who grow up in minority cultures and in societies that embrace many minority cultures.

This intuitive understanding of culture implies that it poses at least two problems for the implementation of the principle of liberty in education. First, because people's conceptions of the good are intimately bound up with their cultures, they can come to have a conception of the good of which they are not consciously aware and to which they have not given their explicit assent. In these circumstances, people become persons but not their own persons; instead they are persons of the culture that they have willy-nilly come to practice. Second, the content of cultures can be inconsistent with liberty, for example, in that a culture may not countenance its members making judgments about their own social roles or in that a culture may be intolerant of the members of other cultures. Thus, these problems, in light of the liberty principle, imply that schools do have an important though hardly exclusive role in the provision of access to culture to enable children to make judgments about and to assent consciously to cultures and the conceptions of the good they partially determine and to discourage the promulgation of cultures that are not consistent with their own and others' personal liberty and, indeed, with the other political values we have developed.

As a result, the liberty principle implies at least two conditions on the cultures to which schools are to provide access. First, those cultures must themselves include a commitment to the value of meaningful personhood for all humans. Otherwise, children may develop a conception of the good that fails to provide for their own or some others' personhood or both. In other words, schools may not provide access to some of the world's cultures, or variants within those cultures, that do not meet this requirement of inclusiveness. Second, those cultures must provide individuals reasonable discretion in determining their places within the cultures. After all, the principle requires not merely that the conception of the good that children develop is socially meaningful but that it is personally meaningful as well. The discretion to adapt one's place in a culture to the purposes the pursuit of which and to the talents the exercise of which one finds significant, for example, is crucial to meeting this requirement. Otherwise, children may develop conceptions of the good that do not enable them to pursue a social role that they find meaningful to themselves. Here, too, the range of cultures to which schools may legitimately give children access is limited. For brevity, I will call the cultures or variants of cultures that meet these two requirements of inclusiveness and role-discretion liberty-consistent cultures. As suggested implicitly, cultures include within them a wide variation in content and orientation, and some of those variants may be liberty-consistent even though other variants are not. Thus, the range of cultures to which schools may provide access is limited to the liberty-consistent variants of world and local cultures.

Even with these two restrictions, however, the number of cultures to which a school may provide children access is undoubtedly very large. Must schools provide access to all of them so that children become their own persons (compare Gutmann 1999)? Certain considerations suggest not. First, the resources needed to make possible children's effective orientation to all of the world's liberty-consistent cultures are undoubtedly beyond capacity of any society. Moreover, unless cultures are actively practiced within a society, they are unlikely to provide children with viable alternatives for their conceptions of the good. Therefore, schools may legitimately restrict the cultures to which they provide access to those liberty-consistent cultures that are actively practiced within that society. Second, these alternatives must be presented to children in a way that makes possible what Bruce Ackerman (1980) has called initial cultural coherence—knowledge of one or two languages and ways of life as they are practiced within the society. In the absence of such cultural coherence, children are unlikely to be able to develop a robust commitment to any conception of the good whatsoever. For, without a rich knowledge of language, an intimate experience of how a culture can frame a potentially worthwhile life, and an early understanding of the ways a culture enables the coordination of social roles in such a life, children are apt to lack the skills and motivation for acquiring a culture and seeking a suitable role within it. Thus, exposure to cultural alternatives is probably meaningful only for children who have already had an effective socialization to one or two initial cultures. For many, perhaps most, children in this country, families and perhaps their surrounding communities provide the initial socialization for this cultural coherence. Presenting children with access to the full range of the world's liberty-consistent cultures, especially early in their lives, may very well interfere with this necessary initial cultural coherence.

As crucial as it may be, however, children's access to culture cannot, according to the principle of personal liberty, be limited only to this initial socialization. On the one hand, the families' cultures may not meet the requirements of a liberty-consistent culture—namely, inclusiveness and individual discretion. On the other, even if these conditions are satisfied by the families' cultures to which children are initially exposed, those cultures may not be presented in such a way that encourages children to respect those who practice different cultures. It need not be that these initial cultures are actively hostile to the members of some other cultures, although that is possible; rather, a natural, primary tendency of the family is to enable children to understand and to live according to the parameters of its culture and to have appropriate facility with its language(s). Families are not likely, with young children especially, to develop an understanding of other cultures in which those cultures are represented as potential sources of their children's conceptions of the good. Yet, such understandings are needed to promote the respect for others and the opportunity to be one's own person that the liberty principle requires. This respect is not mere toleration for what is perceived to be others' outlandish and irrational commitments but a genuine appreciation for what, given appropriate circumstances, could be genuine possibilities for one's own life.

This practical tension in the first principle—between the need for cultural coherence as a foundation for developing children's own conceptions of the good and the achievement of genuine respect among the adherents of different cultures—creates one primary justification for public schooling. That is, schools must both reinforce families' cultures sufficiently to establish a culturally coherent foundation for children's development and provide access to a reasonable variety of realistically available alternatives to the family's culture so that children come to see those alternatives as real possibilities for their own lives. One result of the successful execution of these tasks is that children will emerge with a healthy respect for at least some of the various cultures within their own society. A further possible consequence, however, of providing this access is that a child's emerging conception of the good can be genuinely his or her own. For if children affirm their families' cultures, they will do so in the conviction that those cultures represent personally acceptable alternatives for their own lives, at least from among the cultures to which they have access. And, of course, some children may affirm other cultural possibilities to which they have been given access, possibilities that seem to promise them more satisfactory conceptions of the good than do their families' cultures. But the cultures to which children are exposed need not represent anything like the full range of human possibilities for children's judgments among the possibilities to enable them to be genuinely and thoughtfully implicated in the conceptions of the good that they come to develop and affirm. This process of making a critical commitment to a conception of the good, therefore, need not and indeed should not be understood as a simply choosing among a comprehensive selection of possibilities. For if choice comes into the matter at all, cultures choose us just as much as we choose them. Rather, the process is one of making meaningful and intelligent judgments about and affirmations of some genuine alternative possibilities for our lives.

In this way, children can emerge simultaneously as their own persons and as people who respect others' personhood as representing real possibilities for themselves—both of which are central requirements of the first principle of politically legitimate schooling. Thus, access to a reasonable variety of cultures and cultural variants beyond that of one's family is a necessary condition for the fulfillment of the liberty principle. Because those cultures must represent real possibilities for children's lives, being educated in a monocultural society presents a genuine barrier to the attainment of personal liberty. Diversity in society and schooling, therefore, provides a necessary foundation for such liberty.

Of course, mere exposure by itself to other cultures, no matter how robustly those cultures are represented, does not necessarily imply that children will inevitably emerge as their own persons. In addition to the requirement that such exposure develop in children a reasonable sense of the life possibilities that other cultures present, children must also be helped to develop sensible ways of assessing the significance of those possibilities for their own lives, that is, in light of, for example, the talents the exercise of which they find meaningful. Otherwise, children will make judgments about their conceptions of the good without adequate reference to themselves, and the personhood that emerges in the process will be insufficiently their own. Thus, although an education for personal liberty is significantly oriented to culture, it must also be oriented to knowledge and development of self. This self-understanding requires that children have opportunities to discover, experience, and experiment with their own talents and the sense of worth that may arise from the exercise of those talents. In noting this aspect of legitimate public schooling, two important issues should be addressed at the outset.

First, an education for the development of self is not necessarily an education in selfishness. As Charles Taylor (1989) has explained, self is developed significantly in the context of culture, which largely provides the horizons, the boundaries, of one's personhood. Thus, one's own conception of the self is developed simultaneously with one's acquisition of culture; in other words, to learn to be a Navaho is in part to develop a certain conception of oneself within the parameters of the relevant culture. Whether one assigns a high personal priority to one's own desires or aspirations depends in part on the nature of that culture and the role one comes to have within it. Thus, one can be a self without necessarily being selfish, and the reference to children's selves that their education is to have does not necessarily mean that the conceptions of the good that they develop necessarily emphasize the satisfaction of their self-referential desires.

Second, however, the acquisition of a culture need not entirely determine one's self. Indeed, the personal liberty principle requires there to be a certain critical distance between one's self and one's culture. It implies that in some sense children are to become responsible for the cultures that they affirm as the source of their conceptions of the good. Now, being responsible for one's culture obviously does not mean that one has originated it. Rather, it means that embracing that culture and one's role in it makes good sense to those who do so. The critical distance from one's culture implied by the first principle means, therefore, that one both understands it and has good reasons for affirming it. To be one's own person, it is not enough merely to be socialized to one's culture; one must view it as having a sensible justification considering its fundamental meaning and its relationship to, for example, one's own talents and proclivities.

Of course, one's conception of the good is dependent to a large extent on the opportunities that the cultures to which one is exposed make available. However, it is also to be dependent on the judgments that one makes about and within those opportunities. That is, one must be allowed, even assisted, to form such judgments, and those judgments must, in turn, have an influence on the course of one's development.

An education that is consistent with the first principle, therefore, must allow exposure to multiple local cultures that are inclusive and that allow reasonable discretion about one's social role, but it must simultaneously develop and respect children's emerging judgments within and about those cultures. As already noted, such an education for liberty cannot be the exclusive prerogative of children's families, for the exposure to the multiplicity of liberty-consistent cultures to which children are to have access and the critical distance from those cultures that they are to develop are beyond the usual capabilities and motivations of most families. Nor can such an education be entirely the prerogative of the child. Indeed, the crucial role of culture in the development of personhood makes this second proposition unreasonable. Thus, the authoritative involvement of third parties in children's education for liberty is necessary, especially those people who represent cultures other than that of the family. Of course, this third-party authority cannot be absolute because families make a necessary and legitimate contribution to children's primary enculturation, and children themselves have, as we have seen, a crucial role in the emergence of liberty-consistent personhood through the development and exercise of critical judgment about the cultures to which they have been exposed.

As we will see in chapter six, the precise content and instructional procedures appropriate for this education in culture must be the responsibility and prerogative of local educational authorities operating under certain constraints imposed by central authorities. Here, however, it is useful to indicate in very general terms how such an education might be accomplished. Once the range of cultures to which children are to be given access has been identified, the schools' task is to make those cultures realistic possibilities for children. In providing that access, schools must also enable children to discover and to experience the talents that they possess that may be relevant to those cultures. In accomplishing these tasks, however, schools cannot impose on children authoritative critical judgments of those cultures beyond classifying them as liberty-consistent, for the making of such judgments must be the prerogative of children as they mature in their knowledge of the cultures and themselves. Such knowledge can to some extent be the result of the traditional school curriculum in which children learn about the literature, history, practices, and values of cultures or cultural variants. But it must also involve children's coming to have experiences of the ways of life that those cultures represent. For this purpose, children must interact with people whose lives are structured by those cultures, both as adults and as children. Thus, there must be what might be called an important field component in an appropriate cultural education in which children not only respectfully observe those who live the cultures about which they are learning but also come to experience those lives, at least to some meaningful extent. Exactly, what form this field component of children's cultural education takes will depend on the cultures themselves but also significantly on the imaginative planning of school professionals who have a sympathetic understanding of those cultures.

To sum up, a compulsory public school system, appropriately constituted, is a promising way to achieve this balance of authority among parents, children, and interested third parties. First, such a system must provide children with a sincere appreciation of some liberty-consistent cultures that go beyond that of the family. From the perspective of personal liberty, the curriculum of schools is preeminently a cultural resource, not only a source of, for example, economically, vocationally, or politically relevant learning. Of course, these practical sorts of learning are encompassed by cultures, but a narrow focus on such learning, abstracted from its intellectual and normative basis in the cultures in which it can have social and personal meaning, impedes children's development toward personhood. Moreover, such cultural learning must develop an appreciation of a multiplicity of liberty-consistent cultures as they are practiced in communities to which the child has access. For a focus on a single culture, even if it is different from the child's family culture, amounts to the school's enforcing a particular culture on the child rather than putting him or her in the position of making comparisons of and judgments about multiple live possibilities. Beyond this, the treatment of even multiple cultures as only abstract and alien practices rather than as genuine possibilities for children's own allegiance fails, on the one hand, to present real alternatives to the child's family culture and, on the other, to develop sincere and pervasive attitudes of respect for others' liberty. Second, public authority over children's education must be shared with families, and, as it becomes gradually more feasible, it must be shared eventually with children themselves. Even though there is an unavoidable tension between the family's and the school's goals for children's education, the schools' role is not simply to replace the family culture with a culture that political authorities, however constituted, judge to be superior but, as we have said, to provide alternative cultural resources that children, as they grow to adulthood, can come to understand, view critically, and utilize intelligently as they develop as their own persons.

Personal Liberty and Standards-Based School Reform

Standards-based reform, a contemporary effort to improve student achievement and the schools that students attend, has three basic features. First, central government authorities are to adopt detailed and specific standards for the curriculum and for student learning of it. Second, students at various stages in the schooling process are to be assessed for whether they meet the specified standards. Finally, students and their schools are to be held accountable for their achievement of the standards. In the United States, this strategy has been implemented in a complicated way. During the 1990s, individual states throughout the country adopted various policies informed by this strategy. In 2001, the federal government adopted the No Child Left Behind Act (U.S. Congress 2001) to coordinate, to redefine, and in many ways to supersede these state policies. The federal law requires each state to identify and adopt its own standards for the curriculum and student achievement in reading, mathematics, and science and to test students annually for their achievement of those standards in each of grades 3 through 8. States must also identify a goal for schools' adequate yearly progress (AYP) in meeting these standards, a goal that applies not only to students in the entire school but also to students in various racial, ethnic, family income, English language proficiency, and disability categories. Finally, states must adopt policies for schools that do not meet the standards and the improvement goals, policies that, among other things, must include giving parents in Title I schools the choice to send their children to schools that have met these standards and goals or to receive tutoring and other additional services. The fifty states are meeting these requirements in a variety of ways, but the federal government must approve their methods of doing so.

The exploration of the meaning of personal liberty and the processes by which it can be developed with which this chapter began establishes more fully the grounds for judging the consistency of standards-based school reform with public schooling for personal liberty. In reaching such a judgment, the cultural purpose of the standards must be considered because access to liberty-consistent cultures is, as we have seen, crucial for politically legitimate education for personal liberty. At least three cultural purposes relevant to the development of children toward personal liberty might be operative in the standards that inform standards-based school reform.[1] First, the standards might be understood as providing children with what is necessary for gaining access to any liberty-consistent culture available to them, either the general learning necessary for culture acquisition in their society or the specific learning and experiences necessary to enable them to gain a sense of the meanings of particular cultures and the ways in which those cultures fit with children's talents and proclivities. Second, the standards might be understood as providing children with what is necessary for making critical judgments about the liberty-consistent cultures to which they have access. Third, the standards might be understood as providing a child with learning relevant to one particular culture to which he or she is developing an allegiance. Let us consider the plausibility and legitimacy of each of these purposes for school standards in turn.

If there were a particular configuration of skills and knowledge that all children need to have access to any liberty-consistent culture available in their environment, standards aimed at the development of such skills and knowledge would, indeed, be consistent with, and perhaps required for, an education for personal liberty. For, to fail to make such an education available to any child would deprive him or her of the necessary foundation of personhood. Moreover, such a rationale for the standards might very well justify regional differences in the standards because the foundation for access to the cultures available in one region is different from that of other regions. Thus, the important questions about this cultural purpose of standards are:

- Are in fact any general skills and knowledge necessary for acquiring liberty-consistent cultures in a particular region?
- Do the standards typically adopted by states in fact provide such general skills and knowledge?

Both of these questions are issues, at least in part, of empirical fact.

In an at least minimally literate and numerate society such as that in the United States, it is plausible to maintain that its constituent cultures, including those that are liberty-consistent, are themselves literate and numerate. Here, for example, many of the most central artifacts of our cultures are written. Thus, in this social context at least elementary literacy and numeracy skills are most likely to be implicated in the acquisition of cultures, especially those that are different from one's family culture. Of course, this is not a universal claim about culture since some world cultures are perfectly well transmitted to the illiterate and innumerate, and even in the United States access to significant elements of one's family culture can in most cases be gained without these skills. Rather it is a claim based, on the one hand, on the social facts about our cultures and, on the other, on the importance within an education for liberty of access to multiple cultures. Thus, in U.S. society at least, there is some common learning that likely qualifies as generally necessary for children to gain genuine access to multiple cultures.

The remaining question, then, concerns the extent to which the standards adopted and enforced by the states are focused on and limited to what are the plausibly general skills necessary for culture acquisition in the United States. It is important that the skills and knowledge required by such standards not be any more expansive than what is required for this purpose. For if they are, the standards are in effect transmitting or reinforcing the specific cultures that develop and utilize this nongeneral content. Of course, schools are not to be hostile to particular liberty-consistent cultures in their midst; indeed, it is not the role of government authorities to make judgments among those cultures, for that is to be the prerogative of those who are becoming their own persons. Thus, schools may teach skills relevant to particular liberty-consistent cultures but not others only if the children who develop those nongeneral skills judge them relevant to their emerging conceptions of the good. As noted, however, that standards vary somewhat or even significantly from one state or region to another is not necessarily a sign that the standards of some jurisdictions exceed the minimum required for culture acquisition because that variation can be a legitimate response to the specific configuration of cultures available in different regions.

This is especially true in light of the further requirements of children's gaining meaningful access to multiple liberty-consistent cultures. For such access implies not only children's coming to possess the requisite general skills and knowledge but also, as we have said, children's having experiences of those cultures that will enable them to determine whether those cultures align with their potential talents and proclivities. Again, the purpose of such experiences is both to enable children to gain access to possibilities for their emerging conceptions of the good and to develop a respectful appreciation of cultures other than what turn out to be their own.

Here, regional variations in the liberty-consistent cultures that are available imply concomitant variations in the experiences that are relevant.

The second culturally relevant purpose of standards might be to enable children to make reasonable judgments about the liberty-consistent cultures to which they have access. Beyond knowledge of and experience with liberty-consistent cultures, children are to consider and make sensible personal judgments about the meaning of these cultures for their own lives. Here, too, certain general skills—those of observation, systematic gathering of evidence, interpretation, logical reasoning, self-assessment, and so on—are likely to be pertinent to this purpose, but so, too, are children's knowledge of themselves and even certain attitudes, such as a willingness to postpone judgment until sufficient evidence has accumulated. Of course, such skills, knowledge, and attitudes may be developed in conjunction with the skills needed for culture acquisition and the experience of multiple liberty-consistent cultures, but they go beyond them in that they enable children to gain a critical distance from what they are learning about themselves and the cultures to which they are gaining access. Nor is such learning necessarily neutral to all liberty-consistent cultures. That a culture is inclusive and permits reasonable discretion about one's place within it does not inherently imply that it promotes the sort of reasonable judgments or self-awareness that such critical distance requires. Nevertheless, some patterns of reasoning and forms of interpretation, for example, are justified only within the context of the conception of the good of a particular culture. Thus, for instance, the transcendental deduction may be valid only given Kant's metaphysical assumptions, which are in turn controversial in other metaphysical systems. Similarly, the validity of evidence generated by divine revelation may depend on the acceptance of particular religions. Thus, the standards for and types of reasoning that may be included within an education for personal liberty must be limited to those about which an overlapping consensus exists among the liberty-consistent cultures in a particular jurisdiction, or to what Rawls (1993) termed public reason. Requiring all children to develop any more expansive conception of rationality is in effect to promote particular cultures and conceptions of the good over others, something that is not permitted in a general education for liberty. Of course, once children begin to make commitments about their conceptions of the good, such learning may take place if particular children's conceptions of the good indicate a need for it, but standards that apply to all children cannot require such learning. Thus, some common learning seems necessary for children's making critical judgments about the cultures to which they have access, but such learning must be carefully restricted in scope so that it does not subtly impose on children particular conceptions of the good.

The third culturally relevant purpose of standards might be to enable children to acquire a particular culture to which they are developing a commitment. However, as we have seen, standards, which apply to all children, may not legitimately be based on specific cultures or conceptions of the good, for if they were, they would violate at least some children's personal liberty interests, namely, to develop as their own persons with commitments to conceptions of the good that they find meaningful but that are not reflected in the standards. Once again, learning specific to a particular conception of the good may take place if it results from children's own judgments, but it may not be required of all children, regardless of what their emerging conceptions may be.

Thus, there may be two legitimate purposes of standards relevant to children's personal liberty—the development of general capabilities for access to cultures and for making reasonable personal judgments about them. Standards that are consistent with personal liberties may include those focused on certain skill and knowledge outcomes, but they include, as we have found, particular attitude outcomes as well, such as children's willingness to reach judgments on the basis of adequate evidence. These outcomes must be carefully restricted to what those who hold alternative liberty-consistent conceptions of the good can reasonably accept; otherwise, officials make decisions that illegitimately limit the future liberty of children to become and to live as their own persons. Furthermore, these standards embrace certain educational processes as well as outcomes, namely, they imply that children are to have experiences of multiple cultures, experiences that cannot be reduced to the attainment of specified outcomes, for children are, as a result of these experiences, to reach their own judgments about the fit of various cultures with their talents and proclivities and not to attain pre-determined skills, knowledge, or attitudes. Finally, although, as we have seen, the goal of personal liberty is not necessarily antithetical to the development of standards for schooling, neither are such standards necessarily required to attain that goal. For concluding that certain outcome and process standards do not violate personal liberty is not the same as demonstrating that the adoption and enforcement of such standards is necessary or desirable for an education for personal liberty. For, on the one hand, the processes may occur and the outcomes may develop in a particular social context without the imposition of explicit standards because, for example, there is a widespread understanding among educators and education officials that the requisite outcomes and processes will be included in schools. On the other hand, the regulatory atmosphere established by the enforcement of standards may actually imperil the attainment of the results at which the standards are aimed—by, for instance, discouraging those who have the capabilities to conduct those processes effectively and to achieve those results with students from becoming teachers in the first place.

Without attempting to locate and to assess the complex evidence required to reach a conclusion about the empirical conditions required for justifying the adoption and enforcement of standards suggested by the previous remark, this discussion suggests a pattern of analysis and action for developing standards that are consistent with the requirements of an education for personal liberty. First, education officials ascertain what liberty-consistent cultures exist within their jurisdiction. Next those officials determine what general skills, knowledge, experiences, and attitudes children need to gain access to those cultures and to develop a sympathetic understanding of them. Third, officials determine the skills, knowledge, and attitudes that children need to make good judgments about and among those cultures. Finally, officials design standards for the school curriculum, children's learning, or both that will, in their judgment, enable children to develop such skills, knowledge, and attitudes and to gain such experiences. The critical question is, then, whether politicians and education officials who embrace the recent wave of standards-based school reform have followed procedures that adhere to this pattern.

Apparently, the current practice of standards development, however, is not clearly based on any particular systematic account of the purposes of the curriculum or the schools more generally. Rather, the public became generally concerned about the performance of students in schools and, therefore, about the expectations that schools and teachers have of the students and the adequacy of the efforts they are making toward the fulfillment of those expectations. Political officials and the wider public were persuaded that the performance of both students and schools arose from a lack of specificity in the expectations of them, and that the articulation and promulgation of government standards for children's learning is in part a way of improving that performance. The subjects for which standards were developed were chosen because they are a typical part of the traditional school curriculum, either because states require them or because schools generally offer them in the absence of a state directive. Then groups of teachers, subject matter experts, and other citizens were appointed and asked to specify in some detail which elements of the subjects in question are central to mastering those subjects and to place those elements in a sequence that is both within the grasp of children at various ages and logically necessary for learning subsequent elements in the sequence. Finally, in some jurisdictions and for some of the subjects of special political interest, commercial enterprises are asked to develop tests for children's mastery of these elements of subject matter that are relatively inexpensive and easy to administer. It is crucial to note, however, that in the absence of any explicit commitments about what the curriculum is to accomplish, particularly those relevant to the ways in which the curriculum is to provide for children's future personal liberty, subject matter and student performance standards developed in this way inevitably reflect what political majorities in various jurisdictions find conducive to the ways of life that they find acceptable and desirable, which, in other words, happens to be the learning that will advance the conceptions of the good that are held by the members of the majority cultures in those regions. The subjects selected, the particular content of those subjects, and the nature and extent of children's command of those subjects naturally reflect what the majority find necessary for and conducive to the lives for which they aspire for their children even though cultural minorities are likely to prefer a somewhat different conception of the curriculum that is more conducive to the lives to which they aspire for their children.

It is doubtful that either the officials or the public in those regions are motivated by a grand strategy to enforce a single cultural orientation on the children in public schools. In fact, the piecemeal way in which standards are typically developed belies any such intention. Without, however, a clear specification of what such standards are supposed to accomplish, a commitment from the education leadership or from the grass roots or both, it is inevitable that the unintended, eventual consequence of standards developed in this fashion will be that they reflect the dominant cultural orientation of the populace in the region. As standards expand to the entire school curriculum, a pattern typical of current practice, that dominant culture comes to pervade schooling in the jurisdiction. Of course, democratic politics have a natural tendency to reinforce dominant cultures in schooling in the absence of a conscientious, explicit, and continuing focus on the central purposes of schooling that are contained in the principles of social justice that we have developed, and they have done so by means of many different mechanisms in the past—state specification of subjects to be included in the public school curriculum, statewide textbook adoption, and state school accreditation processes, to name a few. But the current trend toward and method of developing and enforcing curriculum and student performance standards provides localities with much less opportunity to design their schools to be sensitive to local cultures than ever before. Under the current standards regime, therefore, the schools are less and less likely to attend to the cultural diversity that we have seen enables children to develop into their own persons. In other words, political and education officials and the public at large come willy-nilly to act as a democratically authoritarian regime regarding the public schools. Moreover, this can and does happen even though citizens have an intuitive commitment to the four principles of public schooling that we have identified precisely because those principles have not been explicitly and effectively brought to bear on the design of the school standards as they are developed and adopted. For this reason, the current practice of standards development is, in effect, a mechanism by which the schools enforce particular forms of cultural hegemony even though citizens may, and I believe do, hold a commitment to personal liberty. Paradoxically then, the enforcement of dominant cultures by means of standards developed according to the current practice is not consistent with our analysis of an education for personal liberty. For such an education must include a conscientious effort to provide children with meaningful access to multiple liberty-consistent cultures and variants of culture that provide realistic possibilities for their conceptions of the good.

Even though they have not been used to guide practice, the systematic accounts of the purpose of the public school curriculum that are most frequently offered by scholars and other commentators who favor standards-based school reform seem to be of three types—standards should promote economic growth (e.g., Tucker and Codding 1998), cultural literacy (e.g., Hirsch 1987 and 1996), or academic tradition (e.g., Ravitch 2000; Finn, Ravitch, and Fancher 1984). We should now consider whether these theories about the basis of school standards could serve an education for personal liberty any better than the current practice of standards development does.

Schooling to maximize economic growth and thus standards developed for that purpose are patently inconsistent with personal liberty. As we have noted, the usual moral justification of economic growth is the principle of utility, which bids a society to arrange its institutions, including schools, to maximize human happiness. The assertion that happiness, however defined, is the highest human good is, however, a metaphysical commitment of the kind that the liberty principle leaves to individual conscience informed by culture. Thus, to arrange public schooling to maximize economic growth is to enforce on children a particular conception of the good. That enforcement is clearly forbidden by the principle of personal liberty in education. Under a standards regime justified in this way, children simply do not have opportunities to commit themselves to values, occupations, and ways of living that they and others members of their cultures deem worthy and fulfilling but that do not contribute to the continuous growth of the economy. Thus, one prominent theory of the purpose of the curriculum held by some proponents of standards-based reform and accountability is patently incompatible with children's development toward personal liberty.

Cultural literacy as a basis for school standards has far more promise to be consistent with personal liberty. As originally conceived by E.D. Hirsch, cultural literacy is an acquaintance with "the information actually possessed by literate Americans" (Hirsch 1987, xiv). To be sure, Hirsch is not very specific about how to identify "literate Americans"; however, he is not speaking about an exclusive high culture but instead about what average, reasonably well-educated Americans share. Then, on the presumption, reasonable in the United States at least, that some members of all cultures fall within the category of literate Americans, those who share cultural literacy will not necessarily be exclusively participants in a majority culture. Therefore, cultural literacy, conceived in this way, might be tolerably compatible with all cultures embraced by the citizens of the United States.[2] Moreover, school standards based on such cultural literacy will not necessarily involve the enforcement of a dominant culture on all children, as the current practice of standards development does. Nevertheless, what these citizens share is a fortiori not their specific cultural commitments but instead a kind of lingua franca in which they communicate with the members of other cultures for the purposes of commerce, politics, and so on—in other words, a type of cultural pidgin or contact language. Now, an education for personal liberty is not necessarily opposed to the teaching of a cultural pidgin. In fact, in some imaginable circumstances the learning of such a pidgin could possibly be part of the initial stages of an education for liberty in that it might provide those strongly socialized to particular cultures with a way of gaining initial access to other cultures. However, the need for schools to teach such a pidgin assumes that it is not part of what most children acquire by simply living in their society, and in the United States there is good reason to believe that most children are able to communicate with most of their culturally different peers because of their familiarity with pervasive mass media. More important, such an education, should it prove necessary, is definitely not all that an education for personal liberty requires. For such an education must include meaningful access to some of what people with different cultures in a particular society do not share, namely their particular cultural commitments. Thus, school standards based on this conception of cultural literacy do not provide for all the fundamentals of an education for personal liberty—that is, robust knowledge and experience of some cultures beyond that of the family as they are practiced in our society.

Recently, however, Hirsch has come to represent cultural literacy in more robust terms. Rather than a cultural pidgin, he sees the content of cultural literacy now as involving a commitment to an "ecumenical, cosmopolitan" culture designed to replace children's commitments to their families' cultures (Hirsch 1996, 235). He appeals to Immanuel Kant in discussing the intellectual origins of such a cosmopolitan culture. This enhanced view of cultural literacy, thus, espouses an education designed to enforce on children a particular conception of the good, that is, a comprehensive liberalism committed to the fundamental good of autonomy. As we noted about standards based on economic growth, school standards based on Hirsch's current conception of cultural literacy, then, are unquestionably inconsistent with our description of an education for personal liberty because such an education is supposed to leave the decision about whether to abandon children's family culture and if so with which culture or cultural variant to replace it ultimately in the hands of maturing children themselves. Thus, school standards based on cultural literacy either do too little, in that they do not provide meaningful access to other cultures, or too much, in that they are committed to enforcing a particular conception of the good on all children.

Diane Ravitch, for one, recommends standards for schools that reflect the traditional school subjects of literacy and numeracy combined with the disciplines of "history and the sciences, literature and a foreign language" and the conventional values of "honesty, personal responsibility, intellectual curiosity, industry, kindness, empathy, and courage" (Ravitch 2000, 465). Although she makes no mention of the skills, knowledge, or attitudes discussed previously as appropriate to an education for personal liberty, a curriculum based on this very general description could provide much if not all of the requisite education. After all, literacy, numeracy, history, literature, science, and languages are significant vehicles for access to cultures and for the development of critical judgment about them, and thus an education that aims to provide such access will necessarily include some elements or interpretations of those subjects. Thus, simply requiring that these subjects are in some form to be included in the curriculum would allow for the flexibility that local schools would need to tailor the curriculum to the specific cultures that are potentially meaningful alternatives for local students' conceptions of the good as they develop into their own persons and that can provide a concrete basis for the intercultural understanding required by a respect for others' liberty. Thus, Ravitch's list of subjects is not so much a detailed formula for the curriculum as a bare outline of it because the very nature of literature or science, for example, is hotly contested even, or especially, among specialists in those subjects and certainly among the adherents of the various cultures represented in our nation.

However, Ravitch is not satisfied with such a general specification of the subjects of the school curriculum, for that list of subjects is to become the basis for a uniform set of standards for which all children and all schools are to be held accountable. Any particular specification of the content of these subjects that all children in the state or the nation must learn, as required by such uniform standards for the curriculum and student performance, entails that local flexibility to determine the content of those subjects disappears, and so does the opportunity that maturing children have to adapt their school learning to their emerging conceptions of the good. Indeed, when it is the source of state or national standards for student performance, Ravitch's list of mandatory subjects becomes equivalent to the current practice of standards-based reform in that political majorities are expected to determine the particular curricular content, skills, knowledge, and values to which children are to be exposed and with which they must fashion their lives. Ravitch's approach becomes, in other words, a vehicle for the enforcement of dominant cultures, not for access to multiple cultures and children's forming respectful personal judgments about them. In fact, her list of values and virtues that children are to develop discloses that she has formed a judgment about just how the political process for specifying those standards will turn out—that is, as a way of enforcing the values of dominant cultures that not all minority cultures in the country find to be the most significant. Missing from the list of values to be cultivated are, for example, caring, mutual aid, and community solidarity that many cultures place at the center of their ethos.

Thus, although the principle of personal liberty in schooling is not incompatible with the establishment of standards for schools, neither the current practice of standards-based school reform in the United States nor any of the primary theories for the more systematic and principled development of school standards proves to be consistent with the description of an education for personal liberty developed in this chapter. Moreover, this inconsistency does not stem from the principle's attaching a paramount value to individuality or autonomy. As we have seen in the cases of Tucker and Hirsch, the principle of personal liberty as interpreted here is as opposed to a metaphysical form of liberalism—that is, utilitarian or Kantian—as it is to a cultural conservatism represented by the current practice of standards-based reform and by Ravitch.[3] Yet neither does this principle ignore the central role that cultures play in making possible a meaningful personhood for most or all of us. Instead, what it seeks to avoid is the authoritative enforcement of a culture on future citizens no matter what their informed judgment about the acceptability of that culture might turn out to be. It may eventuate that some current or future majority culture proves acceptable to all, but that possible if seemingly unlikely outcome must be the result of judgment, conscience, and persuasion rather than coercion, including the coercion of the young by means of standards established for and implemented by public schools.

CHAPTER FIVE

DEMOCRACY AND EDUCATION: MULTICULTURAL AND CIVIC EDUCATION

Liberal political theory, which political scholars identify as the class of political doctrines that emphasize liberty as a central but not necessarily an exclusive value—that is, a theory like the one being developed in this book—is widely believed to be an inadequate source of civic commitment and thus of civic education (see, e.g., Gutmann 1999 and Sandel 1996).[1] Now, there are many kinds of liberal theory, but all of them have come in for such criticism. Probably the most general complaint about the civic content and potential of these theories arises from their embrace of what is perceived as a pervasive individualism. By placing individual liberty at the apex of political values, liberalism, it is often held, cannot adopt a serious commitment to civic values that supersede individual interests. And because it requires government to be neutral to the various conceptions of the good that its citizens may happen to accept, liberalism is inconsistent with a form of civic education that leads citizens to subordinate their personal conceptions to a robust commitment to the common or collective good. To be sure, liberal theory implies that citizens should mutually respect one another's liberty. But at best, these critics maintain, societies with a strong commitment to personal liberty are simply a loosely regulated competition among citizens to live their lives as they see fit with a grudging toleration for others who are trying to do so as well. However, such societies do not pursue, or even attempt to ascertain, a common conception of the good for all. Without such a conception, it is alleged, they cannot provide for a rich and fulfilling civic life or a compelling civic education.

In this chapter I explore whether the conception of social justice in education outlined in chapters two and three, particularly the principle of democracy in education, provides a response to this critique. That conception, according to the ideas developed by John Rawls in *Political Liberalism* (1993), is politically liberal in that the overlapping consensus specifies that citizens are free within reasonable limits to adopt for themselves the particular conceptions of the good that seem most appropriate to them as individuals and as members of cultures, communities, and other associations. In other words, they can determine for themselves reasonable purposes and ways of living that seem to them to be most meaningful, justified, and fulfilling. For this reason, the members of this society are likely to be in considerable disagreement over their most fundamental moral and intellectual commitments and in particular about the metaphysical premises that justify those commitments. The previous chapter has given us some reason to question the charge of pervasive individualism in this book's conception of social justice in education precisely because of the educational value it attaches to culture, which is hardly an individualistic phenomenon. However, it remains to be seen whether this conception is consistent with a robust and attractive account of civic education.

Civic ideals for this kind of society pose a special problem since one cannot rely on an existing consensus about the moral foundation of those ideals. After all, citizens of this society may, by definition, have widely disparate commitments about their foundational beliefs, depending on the particular conceptions of the good they find satisfactory. Nevertheless, as Rawls suggests and as we have noted in chapters one and two, it may still be possible to create a *political* agreement about the principles that are to inform their larger association by seeking an overlapping consensus in citizens' views about government generally or at least about significant institutions of government, in this case schools.[2]

Now, an overlapping political consensus may appear to be simply the beliefs about government that citizens happen to hold in common. However, such a superficial agreement is not what Rawls has in mind. In *A Theory of Justice* (1999b), for instance, Rawls appeals initially to citizens' intuitions of fairness and their settled convictions of justice. The former is what people in a particular society believe to be necessary conditions for a decision or a choice to be fair, such as the belief that those who make the decision should be impartial. The latter is the specific shared judgments that people hold about the justice or injustice of particular social practices or events, such as the widespread belief that individual freedom is valuable or that discrimination on the basis of irrelevant traits is wrong. Both intuitions of fairness and settled convictions of justice are examples of what people happen to believe. However, Rawls is not satisfied to derive principles of justice from those beliefs alone for the good reason that they almost certainly conflict with one another. A belief in individual freedom, for instance, can contradict the equally widely held belief in equality of opportunity when, say, the exercise of someone's freedom would lead to discriminatory action. Thus, a genuine overlapping consensus requires reflection on, in this instance, just what freedoms and forms of discrimination we have reason to believe are of real political significance and on the relative importance that these more carefully conceived values have when they come into conflict. According to Rawls, the purpose of liberal political thought is, by means of a process of reflective equilibrium, to resolve reasonably these conflicts of belief and practice by ascertaining and prioritizing principles that reflect, to some extent, such intuitions and convictions but that are articulated in a way that eliminates or at least minimizes the apparent conflict.

In reaching that equilibrium, the resulting principles do not necessarily leave the initial conflicting intuitions and convictions entirely or even substantially intact, as we have already seen in deriving the principles of social justice in education. The principles and the priorities among them that result from the process of reflective equilibrium almost certainly will adjust some beliefs to preserve citizens' most central commitments while avoiding some other logical implications of those convictions with which citizens find it most difficult to live. Such principles, and not the raw intuitions themselves, represent for Rawls a genuine overlapping consensus in that they attempt to develop a special sort of consistency among citizens' political beliefs, that is, an equilibrium among their intuitions and convictions achieved by careful reflection on the applications of those intuitions that citizens hold to be inviolable and the applications that are less important to them. Indeed, for Rawls, noticing that some consequences of our beliefs violate other important convictions is a good reason for us to modify or restrict our initial beliefs. The argument for the principles of social justice in education in chapters two and three is an application of this process of reflective equilibrium to Americans' beliefs about public schooling. In this chapter, I consider whether the method of reflective equilibrium toward an overlapping consensus can itself provide not only a way of identifying the principles of social justice in schooling but also an interpretation of the resulting principle that prescribes the role of schools in the education of children about the making of political decisions in their society, namely, the principle of democracy in education.

That principle implies three elements of children's political education. First, it holds that children should be educated for universal participation in their society's political decisions and institutions. Second, it suggests that children should be educated to honor its citizens' basic rights to personal liberty, political liberty, equal opportunity, and economic growth as they participate as adults in the political decision-making process. Third, it holds that children should be educated to participate in a decentralized decision-making system in which localities experiment with alternative social arrangements and learn from other communities' efforts. In the next two sections of this chapter, I consider what a political education that enables children to participate reflectively in the emergence of an overlapping consensus is and whether such education is a reasonable way of fulfilling the three requirements for democracy in education laid out in the second principle.

Multicultural, Historical, and Philosophical Education for Reflective Equilibrium

It is tempting to reach a number of erroneous conclusions about the process of reflective equilibrium involved in reaching an over-lapping consensus. First, it might seem that this process aims at a permanent and immutable state of collective and individual belief about citizens' prerogatives and obligations. However, it is likely that any initial consensus that is achieved through the process of reflective equilibrium will generate civic principles that change how our institutions and social arrangements operate. Such changes inevitably generate new experiences among citizens that in turn help us discover infelicities or even contradictions in our beliefs that were previously obscure. Moreover, our subsequent reflections on those experiences motivate further elaborations and modifications of belief toward new equilibria. Second, it might seem that this process is essentially solitary, involving each citizen in an internal examination of the consistency and acceptability of his or her beliefs and experiences. For two reasons, however, this process is significantly public. One is that the new arrangements to which our temporarily equilibrated beliefs direct us have important public effects in that they naturally evoke responses from not only ourselves but also others, responses that help us understand more fully their meaning and consequences. The other reason is that the assessment of these responses also takes place in conjunction with other citizens who share in the initial overlapping consensus. Therefore, the initial equilibrium produces a new social and ideological milieu that can produce unanticipated consequences and interpretations.

As noted, some of these results can become the motivation for continuing the process of modification and equilibration of belief. Third, and consequently, the process of reflective equilibrium might seem detached from individuals' most central moral commitments, operating entirely in an arena of political negotiation and compromise. However, this putative conclusion radically misrepresents the nature of the process. For the initial intuitions on which the process is based are ineluctably connected to individuals' admittedly diverse personal commitments, that is, their own conceptions of the good. Thus, while those intuitions are shared with others, they are also deeply associated with the various nonpublic beliefs that such a society enables to flourish and that citizens have considerable freedom to adopt and to modify. A change in shared beliefs requires one to consider not only one's reactions to others' responses, actions, and experiences but also the consistency of those modified beliefs with one's own prior foundational commitments. This consideration, in turn, can be the occasion for a revision of one's conception of the good. In this way, the process of reflective equilibrium is continuous and has both public and intensely personal dimensions.

What emerges, then, from the idea of an overlapping consensus that results from reflective equilibrium is a distinctive view of liberal politics. On this view, politics involves significant intellectual and social activity that implicates and influences what citizens believe both about their relationships with other citizens and about themselves. As we have seen, what people believe about themselves and their relationships is modified by a simultaneous process of public and private reasoning. In this process, the political principles that emerge have a moral status because of their connection with what come to be publicly shared and mutually reasonable beliefs and because of their integration with citizens' various conceptions of the good. These principles are, in essence, civic ideals that are not simply facts about people's beliefs or merely a codification of a national civic creed that competes with or displaces citizens' fundamental commitments. Because of the way that they are continuously developed and renewed, those ideals influence and are influenced by private commitments, but because they do not encompass any particular metaphysical foundations, they do not pose a direct challenge to such beliefs. In a real sense, citizens take up the task of seeking and constructing such foundations for themselves and in their own cultural and community associations, but any foundations that they develop do not become part of a societywide public belief system. Of course, an emerging and evolving overlapping consensus certainly influences such private belief systems, but there is no reason to suppose that those systems converge into a single set of metaphysical commitments held by all citizens.[3] Given citizens' initially divergent private beliefs and the commitment of a liberal society to freedom of conscience, in fact, such convergence is unlikely. Thus, an overlapping consensus is compatible both in principle and in fact with a wide diversity of private metaphysical structures of belief and justification. In this way, an overlapping consensus constitutes a set of evolving moral commitments about a nation's civic ideals that is nevertheless harmonious with a wide but changeable variation in citizens' private moralities.

The public education system of such a society can be understood as, in part, a set of government institutions and practices that enable and promote the continual emergence of a reflective overlapping consensus. From this perspective, civic education in public schools is the element of the public education system that undertakes and accomplishes this task for the young. This education is not adequately conceived as simply a vehicle for informing the young about adults' current civic beliefs, for such information is at most only one element of what children need to learn to participate in the development of an overlapping consensus. Nor is such civic education adequately conceived as the enforcement on the young of an authoritative and determinate civic doctrine, for no such doctrine is characteristic of an overlapping consensus because its principles are subject to constant reconsideration and modification. Finally, an adequate civic education is certainly not instruction in a particular metaphysical system of belief, even one with specific civic content or purposes, for such instruction confuses public with private morality. Of course, a fully adequate system of civic education also includes elements that address adults of various ages and in various public roles, but the primary purpose of this chapter is to elaborate in general terms the school-based curriculum and instructional practices appropriate to this conception of civic education.

The aims of the curriculum for such a civic education are relatively straightforward. But in formulating those aims, we must place them in the context of the schools' full contribution to children's moral education. I have argued in chapter four that it is incumbent on a society governed by this conception of social justice in education to provide an education that makes it possible for each child to become his or her own person, an education for personal liberty. As we have seen, such an education includes (1) children's meaningful exposure to liberty-consistent cultures and their associated conceptions of the good, cultures beyond those of the family and immediate community, (2) each child's coming to know about his or her own talents and proclivities, and (3) instruction that enables children to make reasonable judgments about the available conceptions of the good in light of that exposure and knowledge. In this way, public schools contribute to the developing private morality of children without directly determining the substance of that morality. Civic education must operate in conjunction with this education for personal liberty.

Against this background, the curriculum of civic education aims, first, to enable children to learn about the current state of the overlapping consensus—the civic principles of their society and how they derive from widely held intuitions about the relationships and obligations among citizens. Second, such a curriculum must enable children to learn about the meaning and consequences of those principles—how they have been interpreted in the society, the institutions and social practices in which they are instantiated, and the outcomes of those policies and practices, both intended and otherwise. Third, the curriculum must enable children to reflect on the relationship between, on the one hand, those principles and their consequences and, on the other, the overlapping consensus and their developing private moralities. If the curriculum succeeds in achieving these aims of helping children to understand the origin, meaning, public consequences, and personal implications of the society's civic principles, children should emerge from the public school system with the ability to take part as adult citizens in the evolution of the overlapping consensus by means of a process of reflective equilibrium. However, not only must citizens have this ability, but they also must be inclined to make use of it. Finally, an adequate civic education curriculum must, in addition, enable children to see and to appreciate the public purpose and personal meaning of what after all is an intellectually and morally demanding undertaking.

Undoubtedly, many particular configurations of curricular content can enable public schools to achieve these aims of civic education, and the content appropriate to them may vary from one locality to the other, depending on the diverse initial socialization and circumstances of children and the varying histories of the wider societies in which they grow up. In other words, one cannot deduce a specific account of the content or structure of the curriculum from these general aims; they provide only general criteria for constructing and evaluating particular proposals for the curriculum. Moreover, much of the school curriculum that has not traditionally been understood as part of civic education may make an indirect contribution to accomplishing these aims. Language instruction and logical training, for example, provide children with skills that facilitate the requisite learning. In light of these observations, therefore, I analyze only some general aspects of the school curriculum that are relevant to the specifically civic content appropriate to achieving these aims.

I have argued in chapter four that teaching children to understand and to appreciate at least some other cultures in their nation is an important element in education for personal liberty in that, on the one hand, it enables children to consider for themselves conceptions of the good as alternatives to those available in their families and neighborhood communities. Therefore, it expands their freedom to become their own persons rather than persons determined entirely by their immediate social environment. On the other hand, such teaching simultaneously strengthens the entire system of personal liberty by helping children to appreciate others' cultures as real possibilities for their own lives, not just as alien curiosities to be benevolently or perhaps grudgingly tolerated. In addition, teaching about a nation's cultures also makes an important contribution to civic education for an overlapping consensus but for reasons at odds with those most frequently cited in the civic education literature, namely, to facilitate democratic deliberation by helping children understand, anticipate, and negotiate the disagreements that they are likely to encounter in democratic societies (Gutmann and Thompson 1996). After all, teaching about other cultures in their society can also enable children to understand the commonalities as well as the divergences in belief among the members of those cultures. In this way, such teaching can provide children with an understanding of elements of citizens' beliefs that contribute to the current overlapping consensus about political principles and of the various private moral intuitions from which it derives. Thus, the content of an adequate civic education emphasizes whatever commonality of belief that may exist or be possible across cultural differences in addition to the differences themselves.

Combined with instruction that emphasizes our diversity to foster and strengthen personal liberty, the content of the school curriculum, therefore, provides a robust conception of multiculturalism in the society, a conception that expresses both what may unify a nation's citizens and what may divide them. On this view, in teaching children about religion, for example, public schools should not simply endorse an abstract political principle of toleration of others' beliefs while leaving the beliefs to be tolerated unexplored and unexamined. Instead, they should sympathetically and honestly investigate the substantive doctrines of various accessible religions and nonreligious philosophies, including the ways the commitments to toleration and various other public values are understood and justified within them. In an important sense, such cultural learning makes accessible to children knowledge of the habits of thought and the social practices on which a society's overlapping consensus is built and an appreciation of the potential for constructive change implicit in that reality.

Next, although it may be personally useful or even inherently valuable for children to learn about their own and other nations' and localities' histories, the content of history also has a special relevance to civic education for an overlapping consensus. For learning about history presents the opportunity to consider at a remove in time and place the relationships between nations' cultures, their civic ideals, the policies adopted to achieve those ideals, and those policies' results. Especially when the nations or communities under study are liberal societies, the study of history can also reveal the tensions among those four factors and the way the societies have adapted their ideals, policies, and practices in light of those tensions. And when the nation or community under study is one's own, history reveals to children the changing nature of the overlapping consensus and the reasons in the national or community experience for the changes that have taken place in its civic aspirations and ideals. These lessons are crucial for children's gaining an accurate understanding of the nature of an overlapping consensus and for providing them with an appropriate perspective on the tentative status and justifiability of one's own nation's or community's current political principles and policies. Without such a perspective, children might come to regard their nation's or community's commitments to be either absolute or entirely culturally relative. Either of these assumptions actively discourages children from taking the reformulation of a nation's overlapping consensus seriously, for on the first there is seemingly no need to do so, and on the second there is no point in expending one's energy on a matter that is immune from conscious influence. However, the historical education that has just been described is an important corrective to such assumptions. For, first, it is obvious that learning about the changes that have taken place in a nation's or a community's civic ideals and their policy interpretations contradicts the assumption that they are immutable or infallible. But also learning that those changes can be seen as rational, if sometimes mistaken, responses to experience also corrects the assumption that those ideals and policies are nothing but an expression of the majority's untutored cultural preferences.

As one possible example, the history curriculum in schools might consider the social, economic, and religious controversies involved in the debate over slavery before and during the U.S. Civil War and the evolving policy proposals and public policies to which they led. Such a study of the evolving overlapping consensus during this time, the changing public policies in which it was instantiated, the social and economic consequences of those policies, and the various private and public reactions to those consequences can illustrate to children both the tentative nature of civic ideals and the patterns of reasoning employed by citizens at the time to reconcile their private moralities, aspirations, and experiences with those of their fellow citizens.

Admittedly, this curriculum involves a particularly intellectualized view of history, for it entails the perspective that human reason and understanding play a significant role in shaping the national ideals and the events that flow from them. For that reason, it will not be easy for children to master. Nonetheless, it reveals just how profoundly cerebral the task of civic education for an overlapping consensus is.

This quality of the curriculum is equally on display in another crucial and related aspect of its content. Now, an overlapping consensus is the reasonable confluence of popular belief about abstract principles of government and the obligations of citizenship, not merely shared opinions or intuitions about what should be done in particular circumstances. For children to view the rights and duties of citizens as resulting from such principles, the civic education curriculum must also include a philosophical element, in its widest sense. The purpose of this element is to enable children to view their own and others' actions as instances of the application of, to use Immanuel Kant's phrase, maxims of action (Kant 1785/1985). Understanding people's actions as following such general rules implicates and develops children's capacity to abstract from particular actions and to see patterns in them. It may also be one of humans' fundamental logical and moral capacities. Be that as it may, in developing this capacity, one must avoid enforcing Kant's metaphysical doctrines about such maxims—such as, that the only genuinely moral maxims are universal and unconditional—because public education is not to indoctrinate children to accept controversial metaphysical positions. Nevertheless, it is possible to teach children this way of viewing human actions without any particular metaphysical accompaniment. In doing so, one enables children to analyze the actions of governments and their citizens as flowing from general principles, which they can then formulate, reflect on, and perhaps criticize, reinterpret, or reformulate on the basis of their and others' experience and their own private moralities. Indeed, these philosophical abilities can be developed in part in the context of the cultural and history curriculum, as it has been conceived earlier. Children can be invited and encouraged to conceptualize, for example, the principles of government and their rationales that may have emerged from the commitments and circumstances of various social groups during the Civil War era. They can be asked and assisted to understand how these principles changed during the debates over slavery and in response to the events and experiences that they precipitated.

On such a model, children can be enabled to understand current policies and policy outcomes as deriving from both the current overlapping consensus and the tensions that it and its policy consequences create among groups with various cultural orientations and aspirations. For example, they can be invited to consider the phenomena of globalization not only as sources of economic threat and opportunity but also as forces of cultural exchange and challenge that have ramifications for a nation's self-understanding and individuals' and communities' own views of themselves. These abilities to comprehend and to respond to the circumstances of social change are crucial to children's eventual participation in the process of reflective equilibrium as I, following Rawls, have conceived it, for they make it possible to see actions, practices, and policies as serving principles.

This characterization of the content of the civic education curriculum as involving multicultural, historical, and philosophical elements is, no doubt, incomplete. But it demonstrates the kind of analysis necessary for formulating such a curriculum. However, there is one central element of civic education to which the content I have outlined does not necessarily speak, namely, children's motivation to involve themselves in the reflective process through which an overlapping consensus emerges. Achieving this aim, I believe, is less a matter of curricular content than of the instructional procedures through which that content is presented and learned.

Perhaps the key to such motivation is to enable children to explore the connection between the formulation of and adherence to civic principles, on the one hand, and their emerging private moralities, on the other. By this, I do not mean what consequences the principles have for the selfish interests of children, because private moralities, which are substantially based in culture, are not inherently or even usually self-directed. Rather, what I do mean is what consequences these principles have for children's own self-defined interests, which are not necessarily interests in themselves. Nor do I mean that such an exploration should focus only on the teleological consequences of the principles, for children's emerging moralities undoubtedly have deontological as well as teleological components. In short, this exploration involves the connection between the civic principles and what children are coming to believe is right and good.

To accomplish this exploration, it seems necessary to encourage children to assess from their own perspectives the principles that they are discovering in the overlapping consensus. In other words, the teaching about cultures, history, and principles must at some point make room for and facilitate children's reaching their own judgments about the nature and justification of the content of the current overlapping consensus. In part, this means that children must be encouraged to be active and independent in the search for the civic meaning of current governmental and social policies and practices. They must not only be encouraged to formulate hypotheses about such matters, but they must also be encouraged to take seriously the hypotheses of others, including adults and other children. For what they are ultimately seeking is not only their own private interpretations but also an understanding of civic principles that can stand up to public scrutiny. Equally important, they must be encouraged to formulate their own judgments about the adequacy of these principles, judgments based in part on what is publicly known about the principles' content and consequences and in part on what their emerging private moralities make of that content and those consequences. These observations about this aspect of civic education imply a civic education classroom in which children are mutually engaged in the search for the formulation and meaning of their civic ideals and that is respectful of the judgments that children form about them.

Civic Education for an Overlapping Consensus and the Principle of Democracy in Education

With this outline of education for an overlapping consensus on hand, we need now to consider whether such an education meets the requirements of democracy in education established by the second principle developed in chapter two. Initially, we should note some general consequences of this form of civic education that may seem to pose problems for its meeting those requirements. For one thing, it does not teach children specific civic doctrines or prescribe them specific civic duties. Rather, it allows the content of the civic ideals for their society to emerge from their reflection about and their mutual consideration of the cultures, history, and policies of their society. Because of this procedure, it does not directly confront children with an established set of civic values that may or may not conform to their emerging conceptions of the good. Instead, it encourages them to develop civic commitments simultaneously with their emerging personal commitments. In this way, children reach judgments about their preferred personal ideals that are thoughtfully adjusted to the civic practices and principles of the institutions in which they are embedded and to the emerging personal ideals of the other children and the adults whom they encounter. At the same time, the civic ideals and civic duties that they come to embrace are adjusted to what they see as their own possibilities and what makes sense to others in their environment. For these reasons, a civic education for an overlapping consensus encourages children to develop civic and personal commitments that are roughly consistent with those of others around them and with those implicit in the history and institutions of their society. However, this rough consistency does not imply individual children's complete agreement in their civic or personal judgments with either their fellows or with the principles implicit in their current institutions. For, on the one hand, an overlapping consensus about civic principles can be compatible with many different personal conceptions of the good. Thus, while such a consensus may restrict the personal judgments that children reach about their own lives, it does not necessarily determine the details of those judgments, and therefore children can still be free to be their own persons. And, on the other, an overlapping consensus about civic principles is itself subject to continual modification, depending on the policies adopted to implement those principles and citizens' experiences of the consequences of those policies and their implementation.

Now, these general observations about the consequences of an education for an overlapping consensus apparently leave the content of the overlapping consensus itself at least partially indeterminate. Yet, the principle of democracy in education specifies that the content of civic education will include commitments to universal participation, respect for others' rights, local experimentation with social arrangements, and learning from other communities' experiments. However, this disparity is for several reasons more apparent than real. First, this is part of an education that is to be made available to all children in the society. As a result, all children learn the lesson that they are able and motivated to participate in the formulation and evolution of the overlapping consensus. To be sure, such a lesson is more indirect than, for example, that each citizen has an obligation to be actively involved in the political activities whereby the governance of the society is conducted, but it nevertheless is a meaningful form of participation that systematically influences political decisions and that simultaneously leaves it to the judgment of each citizen to what extent more direct forms of political involvement contribute to realizing her or his conception of the good. In a sense, participation in the overlapping consensus is a democracy of thought, principle, and influence rather than a democracy of prescribed political activity, but it is more consistent with a respect for each child's personhood than are a mandatory civic doctrine and externally established civic duties.

Second, this civic education takes place in a society in which, if my analysis of the normative principles of its education is correct, an overlapping consensus already exists about citizens' rights to personal and political liberty and to equal opportunity and economic growth that are consistent with such liberty. To be sure, an education for participation in an overlapping consensus is not a catechism in those rights. Rather, it directs children to identify and to consider the content of the current overlapping consensus. In this society, children who are educated adequately will inevitably discover that the existing overlapping consensus includes commitments to such rights. Although this form of civic education does not drill the children in the content of the existing consensus or indoctrinate them to accept it in precisely the form it currently takes, that consensus and its sources and rationales are important elements in their deliberations. Moreover, the personal conceptions of the good about which they are learning and among which they are forming judgments developed under the same social conditions as the current overlapping consensus. As a result, it is likely but not guaranteed that children's thinking about the overlapping consensus will be significantly influenced by its current state. At the very least, children will take the current consensus's conclusions about citizens' rights and the need to respect them seriously. If they reject those rights as they are currently understood, they will have to do so in conversation and debate with others and in consultation with those in their community who share their own emerging conceptions of the good. This may be the greatest certainty of acceptance that a society that respects the personhood of each can achieve. Besides, if a society's current civic ideals cannot withstand wide critical scrutiny, there is good reason to think that they should change.

Finally, children's investigations of and deliberations about the overlapping consensus take place among their peers and with the adults of their community. This fact clearly provides children with experiences in reflecting on and negotiating over the principles that should guide political decisions that are situated in their local context. Moreover, this experience produces results that enable children to see the value of local experimentation in politics. For inevitably they will discover that the agreements that it is possible to reach with the members of their community are more extensive than those included in the overlapping consensus they discover for the nation as a whole. After all, the diversity of personal and political perspectives in any locality is much more limited than that included in the entire nation, and thus the agreements possible locally are more extensive than those possible in the nation. These tentative agreements about what is desirable and feasible in the local community provide the basis for local and regional political experiments that are consistent with the national overlapping consensus but that go beyond what the national population is as yet willing to embrace. Of course, many of these experiments may fail to achieve results that satisfy local aspirations. Many of those that succeed locally may not seem acceptable to the citizens of the nation. But some will prove successful and attractive enough to persuade the members of other communities and perhaps even the nation eventually to adopt them. In this way, local experimentation and the requisite skills and attitudes among local citizens have the potential to change the overlapping consensus of the nation. Even when this does not occur, a civic education for an overlapping consensus that is unfettered by undue restrictions imposed by larger jurisdictions of government enables citizens of local communities to create political arrangements that promise to satisfy the conceptions of the good held by members of those communities. As with respect for citizens' rights, there are risks that attach to such a localized civic education, in particular the possibility that the local consensuses that emerge may violate the principles included in the overlapping national consensus. As noted, in some instances these departures from national norms may provide the basis for constructive substantive change in those norms. In others, they may simply indicate a need for national norms to become more tolerant of certain local variations. But in the most extreme cases, they may reflect a locality's unwillingness to accept the overlapping consensus of the nation, and these eventualities may necessitate the enforcement of national norms on deviant communities. Nevertheless, the potential of political decentralization of schools and other government activities both to satisfy local aspirations and to provide a source of fruitful change in national norms make these risks worth taking.

In sum, a civic education for an overlapping consensus does meet the requirements of the principle of democracy in education in that it provides a universal form of political participation, respect for the rights included in the overlapping consensus, and experience with local deliberation about and experimentation with political solutions to social problems. However, it does not meet these requirements in the most obvious way, that is, by enforcing on children a particular political doctrine. Nevertheless, this kind of civic education is far more consistent with the spirit and intention of the principles of justice in education as a whole than a more direct socialization to fixed civic ideals would be. First, it takes the judgments of children seriously, and, in that way, it respects both their personal and their political liberty. Second, it is more democratic since it subjects not only the actions of government but also the society's civic ideals to the judgment of its citizens.

Civic Education beyond the Overlapping Consensus

This analysis implies that Rawls's conceptions of an overlapping consensus and of the process of reflective equilibrium from which that consensus emerges suggest a civic education that aligns with the requirements of the principle of democracy in education. Moreover, this civic education is more robust and demanding than critics of liberal theory suppose is possible. While such an education takes note of what citizens happen to believe about themselves and their fellow citizens, it enables and encourages children to reach public and private moral judgments about those beliefs. Moreover, the judgments that children reach are not simply the application of an established and official civic doctrine but are instead the result of a thoughtful analysis of the public meaning of civic principles and of an assessment of those principles' capability of meeting the requirements of children's emerging private moralities. And because of that analysis and assessment, children have self- and public-referential reasons to engage honestly and actively with their society's civic ideals and to take seriously the rights and responsibilities of citizenship.

On this account, then, a civic education for an overlapping consensus contributes simultaneously to the construction of the self and to the construction of one's society, and it does so interactively, so that the emerging self is not simply a result of internalizing norms that are supplied from without, as a civic religion might imply. Neither, however, are one's social ideals simply a matter of applying one's own conception of the good to the principles, policies, and institutions of society, as one's private morality might bid one to do. In this way, civic education can be a complex kind of public moral education in which students learn from and teach themselves and others. And contrary to the claims of deliberative democrats (Gutmann 1999) and communitarians (Bellah, Madsen, Sullivan, Swidler, and Tipton 1991), political liberalism makes possible an attractive if demanding civic education that is much more public than they believe possible in a liberal society even though it is not based on or does not seek a specific conception of the collective good. Rather than an irresistibly privatizing civic morality, political liberalism implies, as we have seen, an education for involvement in public dialogue about civic values that nevertheless does not require that the demands of private morality are ignored or eclipsed.

In recent years, several other liberal theorists have also responded to the charge that civic education in a politically liberal society cannot be sufficiently robust to enable children to develop a strong civic commitment to such a society. Their deliberations reach many conclusions similar to those developed so far in this chapter, particularly relating to the need for dialogue with and among children about civic matters in public schools. However, these theorists recommend additional education policies to support the civic purposes of liberal societies, including policies for teaching children to value their own and others' autonomy, policies expressing caution about schools' responding to the demands for accommodation made by parents who do not accept some of the liberal society's basic commitments, and policies for developing a sense of community in the public schools. We should consider whether such policies are a necessary complement to an education for the overlapping consensus that I have outlined here.

Eamonn Callan (1996 and 1997), for example, has argued that by examining the implications of Rawls's ideas for civic education one is led to conclude that political liberalism is actually a form of what Callan calls ethical liberalism, a partially comprehensive doctrine that entails metaphysical commitments, particularly to personal autonomy, of the kind that Rawls thought he could avoid and that, therefore, pose a direct challenge to some of the personal conceptions of the good that children in a liberal society might develop. Specifically, Callan maintains that liberal civic education requires children to recognize that public reason is not powerful enough to entail any particular conception of the good to the exclusion of others. As I read Callan, to recognize the limits of public reason, children must establish a critical distance between themselves and their own emerging conceptions of the good, and this critical distance implies that children must come effectively to hold a metaphysical commitment to an even higher good, namely, to personal autonomy.

However, the scheme of civic education that I have suggested does not necessarily require children to confront the issue of the limits of public reason explicitly or even indirectly. On the contrary, the process of public dialogue in schools about the overlapping consensus and the accompanying continuous self-examination of children's own emerging conceptions of the good does not necessitate that children occupy a critical space from which any and all reasonable conceptions of the good can be viewed and judged. After all, even when it concerns matters of national or international interest, this dialogue and self-examination take place locally, and the adjustments in the overlapping consensus and children's emerging personal conceptions of the good that occur do not require children to develop anything like an objective "view from nowhere" in which they come to have a distinctive understanding of the limits of human reason. In fact, the critical perspectives that children develop are distinctly a view from the "somewhere" in which they are already situated, that is, the community in which they live and the conceptions of the good available in that community with which they are fashioning their own conceptions. Thus, such a civic education does not require children to view or to value themselves, much less human selves more generally, as radically detached from their family, social, community, or cultural contexts. Therefore, I suggest, such a civic education does not depend on or promote a metaphysical commitment to individual autonomy. Of course, some children may judge such a metaphysical commitment to be an attractive element of their conceptions of the good, but neither the education for personal liberty nor the civic education I have outlined demands that they do so.

In his effort to represent the civic education implications of political liberalism as substantive and muscular, Stephen Macedo (1995 and 2000) emphasizes the potential for educational confrontations between a politically liberal state and some parents' conceptions of the good, particularly the religious aspects of those conceptions. In his analysis of the court decision in *Mozert v. Hawkins* (1987), for example, Macedo argues that a politically liberal society has no particular reason to accommodate the position of religious parents who requested the exclusion of their children from a reading program that they believe contradicts the teaching of their church and who threatened to withdraw their children from the public schools if their request was not satisfied. There is no doubt that a politically liberal state has reason to resist some parents' demands about the education of their children, but Macedo, I believe, is not sufficiently alert to the crucial role that cultural and religious differences have in making an education for personal liberty possible and these differences' contributions to an education for an overlapping consensus. As a result, Macedo does not take John Rawls's (1999b, 190–194) discussion of the toleration of the intolerant seriously enough. In that discussion, Rawls concludes that a liberal society should accommodate even citizens who reject important elements of the overlapping consensus as long as they do not pose a real and present danger to the liberal regime. Although Rawls does not apply his conclusion to the educational context, that conclusion is particularly relevant to schools. First, by accommodating such intolerant parents in public schools, their children continue to have the opportunity to receive some important elements of an education for personal liberty, in particular, an exposure to living examples of those who abide by more tolerant conceptions of the good, and other children learn an equally important lesson in liberal toleration. Second, by enabling those children to remain in public schools, they and other children gain an opportunity through the civic aspects of their education to deliberate about the extent to which such views can and should be accommodated by the overlapping political consensus. Thus, the toleration of the intolerant is not just a moral duty of the liberal state, but it also can have significant educational consequences as well.

Kenneth Strike (1994) argues that John Rawls's politically liberal society requires, like any society, an education in what he calls an authoritative, shared civic language to promote social cohesion, but in the context of a politically liberal society such a language or an education in it runs the risks, on the one hand, of being so robust that it renders the society inconsistent with its pretensions of neutrality toward its citizens' conceptions of the good or, on the other, so emaciated that it cannot form the basis of an education in content beyond the civic that its students can recognize as worthwhile. He suggests, however, that there exists a third possibility for civic education that avoids these risks, namely, a dialogue between the members of the society's and the schools' diverse communities about the nature and justification of their shared civic language. In many ways, I find this dialogue to be significantly compatible with the education for participation in the evolving overlapping consensus that I have described. Beyond this civic dialogue, however, Strike (2004) now further suggests that without explicit attention to the community dimensions of the institution, schools in a liberal society can foster alienation and disengagement among students for many of the same reasons that he worried earlier about the effects of an emaciated civic language. Of course, the civic education I have described is not necessarily opposed to measures to enhance the community life of schools, but I believe that Strike does not fully appreciate the potential of liberal civic education to resist the alienation about which he is concerned. Just as I pointed out in response to Callan, the dialogue to which such civic education leads takes place locally, that is, in the contexts both of the communities in which children find themselves and of children's emerging conceptions of the good in those communities. Therefore, the dialogue does not enforce, in effect, a commitment to personal autonomy on children, but it also encourages children to understand the overlapping consensus in terms that relate to what is familiar to them and what are emerging as their own commitments to meaningful lives. Thus, the civic framework of public schooling in a politically liberal society does not invite children to become alienated from their own communities or what they themselves are coming to value, and it, therefore, does not necessarily encourage students to become disengaged from the non-civic elements of their education by depriving those elements of the sources of personal and communal meaning. In fact, we will reinforce the strength of this conclusion in the next chapter's consideration of the way that schooling can legitimately be differentiated in response to children's emerging conceptions of the good. Now, in light of this analysis, whether the community life of schools needs to be strengthened beyond what becomes automatically available by means of an education for personal liberty and for an overlapping consensus depends on the communal nature of the various conceptions of the good that are available in their communities and that children come to embrace. For schools to emphasize the value of community beyond what is implicit in children's own cultures and their own emerging conceptions of the good is a form of cultural imposition to which this conception of social justice in education is opposed.

A consideration of Callan, Macedo, and Strike has added important clarifications to the discussion of civic education for an overlapping consensus. In particular, these theorists have helped us understand how an education for democracy combined with an education for personal liberty work together to enable children to enjoy both meaningful individual freedom and a sense of civic responsibility in their local contexts and in the nation as a whole. However, such an outcome does not require children's education to place an independent value on either personal autonomy or community or to take an excessively confrontational stance toward those who do not share in the overlapping consensus.

CHAPTER SIX

EQUAL OPPORTUNITY AND
EDUCATION: CONTROL OF SCHOOLS

It might be tempting to conclude from the discussion of children's education thus far that control of schools should reside entirely with the local community or even with the individual school and its constituents. After all, the content of schooling that is effective in providing for the personhood and thus the personal liberty of children differs with local and regional variations in cultures because children's conceptions of the good are significantly shaped by those sources. Moreover, the education for an overlapping consensus as described in the previous chapter must both avoid the imposition of a specific national civic doctrine on children and prepare them for a form of political decision making that is informed by the insights and ambitions of local communities in light of their cultural values and historical idiosyncrasies. What cannot be permitted to vary, however, is the commitment of schools to the principles of personal liberty and democracy.

American history provides us with ample evidence that local and state control of schools sometimes leads to the violation of these commitments. In Oregon in the 1920s, for example, a law was passed that forbade children from attending parochial schools, with the intention of depriving Roman Catholic children of the chance to receive an education that was consistent with their families' religious beliefs (*Pierce v. Society of Sisters* 1925). A Pennsylvania law required schools to read students passages from the Bible each day, with the purpose of promoting a specific version of Christianity as opposed to other interpretations and other religions (*Abington v. Schempp* 1963). There have been many efforts by parents or school boards to restrict the content studied in classes or the holdings of school libraries to prevent children's learning about beliefs and their attendant ways of life that they or the majority of the members of the local community find undesirable (e.g., *Todd v. Rochester* 1972). Finally, some schools have attempted to restrain students from expressing controversial political views (e.g., *Tinker v. Des Moines* 1969).

Now, it is important to be clear about why policies like these contravene the principles. It is not that schools' curricula must exclude all religious and controversial political content. Quite the contrary, schools must give children access to a reasonable diversity of religious and political views that make available to them live possibilities for their own conceptions of the good and that give them the opportunity to develop a respectful regard for the religious and political perspectives they encounter even if they choose not to make some of those views an element of their own conceptions. Thus, the problem with the Oregon and Pennsylvania laws is that they fostered too narrow an understanding of the religious views that were available and respectable. Schools do not have an obligation to present all possible or actual conceptions of the good, but they must determine the range of conceptions that the curriculum does present on the grounds that they are viable, liberty-consistent possibilities for children's lives and not simply because some members of the community do or do not hold them or do or do not approve of them. Thus, the problem in Rochester, where parents wanted the curriculum to change because they found it offensive to their religious beliefs and similar cases is that curricular or content restrictions were sought for inappropriate reasons. Similarly, in Des Moines the restriction on political expression was determined by the popularity of the views expressed and not by a consideration of whether those views had a potentially salutary effect on children's learning and deliberation about the evolving overlapping consensus in their community or the society at large. These risks of giving localities control of their children's education seem to argue for state and national control of schools in the interests of enforcing the first two principles of social justice in education.

These local and regional divergences from the principles of justice in education have been particularly troubling in the case of equality of opportunity. The shameful history of legal and de facto segregation of and discrimination against racial, ethnic, and disability groups in American schools that the courts struck down in a series of cases beginning with *Brown v. Board of Education* (1954) demonstrates that state law and local practice can be formidable barriers to the achievement of equal educational opportunity.

As with liberty and democracy, we must be precise about the way that these policies contravene the principle of equal opportunity in schooling as we have formulated it. Clearly, prohibiting some children from attending public schools, as was once the practice with many children with disabilities (see *Mills v. Board of Education* 1972), deprives them of the chance to develop their abilities that is provided to others who are allowed to attend. In this way, the exclusion of some children from public education is a patent violation of the principle of equal opportunity as well as the principles of personal liberty and democracy. However, most of these policies did not exclude children altogether from attending public education. Nevertheless, providing education in a manner to which some children do not have access, as was the practice with many children who did not speak English (see *Lau v. Nichols* 1974), effectively excludes those children from the instruction that others receive and thereby comparatively diminishes their chances of development. Thus, effective exclusion even though attendance is permitted is also a violation of the principle of equal opportunity and other principles as well.

Some of these policies, such as racial segregation and other blatant and more subtle forms of discrimination permit attendance and access to the instruction provided. These policies cannot be faulted for the mere fact that they provide access to differential instruction. After all, the principle of equal opportunity permits, indeed requires, differential instruction for at least two reasons. First, if children have different emerging conceptions of the good, they may seek to develop different abilities, in which case they are likely to need instruction that differs in form and content. Second, even if they have similar conceptions of the good, individual children may have experienced more or fewer social disadvantages, which imply that the more disadvantaged may need different amounts or kinds of instruction to enable them to overcome those disadvantages in developing the particular talents that are relevant to their conceptions. What make segregation and other discriminatory policies unjust is that they inherently require the determination of the education that children receive on grounds other than what their conceptions of the good and their educational backgrounds require. Thus, segregation laws prescribe a differential determination of the schools children are to attend on the sole grounds of their membership in different racial or ethnic groups rather than the different aspirations that they may have or the specific barriers that they face as they attempt to realize those aspirations, in contrast to, for example, their freely choosing schools or programs that have different curricular emphases that are adapted to realizing those aspirations or to overcoming those barriers. These discriminatory attendance policies, therefore, violate the principle of equal educational opportunity precisely because they make educational decisions for inappropriate and thus unjust reasons. To echo the Supreme Court in *Brown*, segregation and other forms of discriminatory placement are inherently wrong even if the schools or programs to which children are assigned offer the same instruction because children's conceptions of the good or their disadvantages in pursuing those conceptions are not considered in making the assignment of schools or instructional programs. Of course, if, as is usually the case, racially or ethnically segregated schools offer systematically dissimilar educational programs,

programs that differ either in available content or in the quality of the instruction provided, the initial injustice of segregation is further compounded. Similarly, discriminatory placement of children into educational programs within schools is usually intended to provide those children with differential instruction, again for reasons that contravene the principle of equal educational opportunity. Under such policies, children simply cannot have an equal chance to develop the differential abilities they find to be important because of their diverse emerging conceptions of the good. Here, too, an apparent need for national control of schools also emerges from a consideration of the history and current practice of state and local control.

Before concluding, however, that the principles of justice in education imply thoroughly centralized control of public schooling, we should note that such control poses a number of related problems, depending on the way that it is exercised. If, for example, central control of schools is taken to mean that all schools are to have the same mandatory curriculum at all levels, schools would be unable to deliver an education that is sensitive to local and regional variations in culture that the arguments about the principles of personal liberty and democracy suggest is necessary. Thus, a uniform national curriculum, or even one that is uniform within a particular state, would not permit the variation in content and instruction that a respect for children's emerging personhood and citizenship requires. Moreover, such curricular uniformity enforces on children a single authoritative definition of success, most likely that of the dominant culture in the nation, that the discussion of equal opportunity in chapter three concluded is inappropriate. After all, children's differential emerging personal conceptions of the good imply differential understandings of what success in life amounts to, and the pathways to achieving these different accounts of success will be as various as the conceptions themselves. Undoubtedly, these accounts of success will be shared with some or perhaps many others in the society, and even widely divergent accounts may necessitate some common learning, but a completely uniform curriculum for all is an unlikely vehicle for the equal support of all children's diverse aspirations that the principle of equal opportunity in schooling implies.

Another possibility for central control of the curriculum is that it would be differentiated according to the various predicted needs of the economy and the wider society, and children would be assigned to the several curricula according to their ability to master them and, thus, to fulfill those social needs. To be sure, this policy would allow for significant variations in children's school experience and in the definitions of success to which it leads. However, this strategy would enforce on children specific definitions of success that may or may not be related to their particular emerging conceptions of the good. Even worse, it would constrain many, perhaps most, children to develop conceptions of the good that are determined in advance of their coming to make personal judgments about good lives, and in this way children would emerge not as their own persons but rather as the society's as a whole. Thus, this policy would violate the principles of both personal liberty and equality of opportunity in that children's judgments about their conceptions of the good and about what they deem appropriate for realizing them are neglected. Even if children were permitted to choose among the various curricula determined by society's needs, there is no guarantee that any of those curricula would match particular children's developing judgments of the good, and both of these mechanisms of enforcement and constraint would still be operative.

In chapter three, we considered yet another possibility for a partially centralized curriculum, namely, a uniform definition of a threshold of outcomes that all children are supposed to attain, coupled with children's freedom to pursue other outcomes once they attain the threshold. We also rejected that possibility for a number of reasons. The most compelling was that, ironically, such a strategy is a formula for providing socially marginalized children with no educational assistance in developing a comparative social and economic advantage while enabling other less marginalized children to develop such an advantage. Thus, such a centrally determined threshold will, at the very least, perpetuate the advantages of children from relatively privileged families and may even exacerbate them. This consequence is clearly at odds with the principle of equal opportunity, which requires schools to compensate for any relative social disadvantages that children may have in developing their conceptions of the good and the abilities to pursue them.

Amy Gutmann (1999) provides a different interpretation of the learning threshold that all children are to attain, namely, the skills and knowledge that will permit children as adults to participate in the democratic political process in their particular society. Kenneth R. Howe (1997) has observed that if this learning for democratic participation is determined independently of the self-defined needs of marginalized children, schools are likely to be a source of their oppression. Therefore, he has extended Gutmann's interpretation further to suggest that what children are to learn to participate effectively in the democratic process should itself be determined by means of a participatory process in which children and their cultural, racial, ethnic, gender, and disability perspectives, especially those of marginalized children, are effectively represented. Now, I am sympathetic with what both Gutmann and Howe have to say about the preparation of children for democratic participation, and indeed these considerations are to some extent important to what and how children are to learn to fulfill the principle of democracy in education discussed in chapter five. However, this interpretation simply fails to recognize the differential learning that children need to enable them to be their own persons and to develop their comparative advantages in that light. Thus, this interpretation, even with Howe's qualifications, does not adequately capture the underlying meaning and rationale of either the principle of personal liberty or that of equal opportunity in education.

These various mechanisms for central control over the curriculum may not be the only possibilities, but they are the ones most frequently mentioned in contemporary philosophical and policy discussions, and they all fail to meet at least some of the requirements of the principles of personal liberty, democracy, and equal opportunity in education. Moreover, any other means of central control of the curriculum that prescribes a uniform content that is not flexible enough to accommodate both the various learning needs of children who are developing diverse conceptions of the good and the wide range and constellation of cultures in particular localities and thus of children's emerging conceptions of the good is doomed to violate these principles. At most, such a central mandate could specify an extremely basic curriculum in literacy and numeracy, say, for as we have noted the many cultures and conceptions of the good in this country all demand those basic skills, but even such a basic curriculum must be adapted to the particular cultural contexts of each community and the specific developing aspirations of students in it. That is, the curricular content by means of which children come to be literate and numerate must be attuned to their backgrounds and developing personal judgments.

What is more, it is extremely unlikely that communities would fail to see the value of and to adopt such a core curriculum. Therefore, while there is no particular harm in such a central mandate for a basic curriculum, little if anything is to be gained by such a requirement. Moreover, a locality that controls its own curriculum is likely to make appropriate adaptations of curriculum and instruction to the needs of local children. The real problems, as our history tells us, lie with communities' effective exclusion of some children from the curriculum they offer, their providing discriminatory access to that curriculum, and their failure to adopt a curriculum beyond the basic core that is adequately broad and flexible enough to facilitate the emerging ambitions of all local children. Preventing these problems, then, is the important role for central authorities in governing schools. Thus, the principles of social justice in education in our society imply a division of responsibility between central and local authorities in the governance of schools to meet the likely threats to the realization of those principles.

Central Authorities' Responsibilities in School Governance

The common wisdom of recent school reformers about the division of responsibilities between central and local authorities in schooling is that central authorities should specify the "what" of education, that is, its content and expected outcomes, and local authorities should decide the "how" of children's instruction, that is the administrative, organizational, and teaching methods that will be effective in delivering that content and achieving those outcomes (e.g., Fuhrman and Odden 2001). On this account, central authorities should determine which subjects should be taught and the expectations for students' performance in those subjects, or, in the language of these reformers, the educational standards. The task for localities, then, is to design and select the specific strategies of instruction that will enable their children to meet these standards in light of the children's cultural backgrounds and their previous experience. Our discussion thus far, however, implies that this straightforward division of responsibilities must be mistaken. For we have found that central authority over the curriculum is probably not necessary if it is restricted to the basic skills that all children need whatever their conceptions of the good and their cultural backgrounds turn out to be, and it is illegitimate if it goes beyond such basic skills to specify an education either that conflicts with children's emerging conceptions or, even worse, that determines those conceptions in advance of their having the opportunity to make judgments about them. On this view, moreover, children's background and experience are seen as simply either barriers to or facilitators of their achievement of common learning. But as we have found, that background is an important resource to children as they develop their conceptions of the good, and children's experiences are partial sources of the judgments they make about what life is worth living, judgments that schools must enable children to make and that they must respect as children are making them.

In large part, this recommended division of responsibility goes wrong because it neglects the "why" of children's learning, which is specified by the four principles of social justice in education. First, children are to learn to be their own persons. Second, they are to learn to become participants in the continuous formulation of the overlapping consensus. Third, they are to learn to develop their competitive social advantages guided by their own aspirations. Fourth, they are to learn in what way the economy plays a role in the pursuit of their conceptions of the good. These purposes of schooling, as they represent the current overlapping consensus about the education of the young, define responsibilities on which the exercise of central authority over schools should be based. Thus, central authorities first and foremost should establish a legal framework in which these purposes of education are established, enforced, and facilitated as the primary responsibilities of the public school system. And the system of accountability for public schools should assess primarily whether localities are achieving these purposes, not, for example, whether all students are achieving at specified levels in particular subjects.

Because of these purposes, the standards-based reformers' view of the division of authority also goes wrong in that it ignores the legitimate roles that central authorities have in holding schools accountable for the "how" of schooling and that local authorities have in determining the "what." As we have found, the uniformity in content and outcomes beyond very basic skills that the exercise of central authority may establish is entirely out of place in a system of public schools designed to achieve these purposes. Thus, to a significant extent, the control of the content of education must be the responsibility of local authorities, and central authorities' responsibilities beyond establishing and monitoring the achievement of these purposes must lie largely in their concern over the process rather than the outcomes of schooling. In the next section, we will consider the responsibilities of local authorities in some detail, but here we should consider how these primary purposes of public schooling give central authorities control over aspects of the process and procedures according to which schooling is conducted.

One significant dimension of the educational process that falls outside the legitimate responsibility of central authorities is the specific set of instructional procedures to be used in schools and classrooms. After all, the types and patterns of instruction need to be adjusted to the cultural backgrounds and aspirations of students as they develop and pursue their conceptions of the good. Therefore, centralized uniformity in methods of instruction is inconsistent with the principles. However, there are many other aspects of school procedure that are legitimately of interest to central authorities.

A basic procedural concern of central authorities lies with the requirement and enforcement of compulsory school attendance. As argued in chapter two, such compulsory attendance does not violate the personal liberty of either parents or their children because, on the one hand, parents' liberty interests do not include the right to prevent children from developing as their own persons or to prevent others in the society from attempting to influence the conceptions of the good that children develop. On the other hand, young children's personhood is not developed sufficiently to establish the fulfillment of their current preferences as a genuine liberty interest. Indeed, children's schooling, if conducted appropriately, helps them to gain the experience and to develop the judgment necessary to ensure that the conceptions of the good that emerge are sufficiently their own to establish such an interest. Moreover, the education for personal liberty and democracy that we have outlined in chapters four and five help ensure that children's emerging claims on personal liberty will be recognized in ways that are both respectful of others' similar claims and that take into account others' deliberative judgments about the political and social conditions necessary for them to pursue their diverse interests cooperatively.

Thus, central authorities have a legitimate concern that all children attend public schools so that they receive instruction that satisfies the liberty interests of children and other citizens and that, as we shall consider next, satisfies both groups' interests in the fair conduct of differentiated education.

The principle of equal opportunity in education, however, also recognizes that the pursuit of these interests also has a competitive as well as a cooperative dimension. It is important to understand just what the principles of justice in education imply about what legitimate competition means and does not mean for schooling. Legitimate competition does not imply that one child's becoming her own person and a citizen may diminish another's attaining those conditions. Personhood and citizenship are to be fixed points in every child's education. But legitimate competition in a society governed by the four principles can mean that one person's interests in who becomes a member of the clergy, a nurse, or a software engineer, for example, can possibly affect others' chances to do so. For these chances depend in part on the aspirations and the realized talents of particular children, which to a significant extent are functions of their own developing conceptions of the good. If I happen to aspire to become a nurse, but I am not able to master or do want to work hard at finding out about people's complicated illnesses and the appropriate treatments, I am less likely to be able to realize my aspiration than those who have the appropriate abilities and motivation necessary for completing nurse preparation. Thus, these chances vary with the extent to which a child possesses the requisite aptitudes and motivations in comparison with others. Just as important, one's occupying these roles can have crucial consequences for the fulfillment of others' conceptions of the good. Becoming a competent and caring nurse, for example, can positively affect the chances that both the consumers of one's nursing services and one's nursing colleagues have to fulfill their conceptions of the good and vice versa. Thus, one's opportunities to take on these roles also depend on the extent and the nature of the demand for those roles created by others' conceptions of the good. For example, if others aspire to lead a spiritual life, one's chances of becoming a member of the clergy are thereby enhanced. Thus, one's chances to occupy these social and economic roles depend on one's aspirations, one's achievement, and the social demand for those roles. As children's evolving conceptions and their subsequent pursuits of those conceptions affect the prevailing configuration of these factors, they also affect these chances.

These remarks, moreover, contain not just a description of the facts about competition in society but also a specification of the norms that should govern such competition. The principles of social justice in education imply that the competition for these social roles ought to depend on precisely these factors because they all reflect a respect for people's various conceptions of the good. That is, no one in a society governed by these principles has a right to force another to be a nurse, for instance, if his conception of the good does not call for it. Further, no one has the right to demand employment as a software engineer if she has not mastered, for example, the skills that others' conceptions of the good require in that role. And no one has a right to demand such employment if others' conceptions of the good simply do not require the activities that the members of these occupations perform.

In light of this understanding of the nature and justification of legitimate competition over social roles, educational authorities have specific responsibilities in preparing children for those roles. As we have seen, these authorities do not make the final decisions about the allocation of children to this preparation, for children's own aspirations are to play a central part in that allocation. Rather, the role of education authorities is to provide children with fair access to that preparation. Part of that function is, of course, to give children effective instruction in any basic skills that are prerequisite for all of these social roles. Further, we have seen that education for personhood includes the development of children's knowledge about their own talents and the judgment necessary to understand how those talents are or are not relevant to the various conceptions of the good that they might find fulfilling. In addition, such knowledge and judgment play a crucial role in the fair access to social roles in that they give children an understanding of their prospects for developing the abilities required by those social roles. Beyond this, however, children must understand the implications of others' expectations about these roles, including the precise level of ability that those expectations require, the work needed to develop those abilities, and the extent of the social demand for them. With these combined understandings of themselves and their society, children are in a position to determine for themselves honestly where their comparative advantages in the society lie. Finally, the instruction to which they have access must be responsive to these determinations. First, as children gradually make these judgments, schools have a responsibility to make available to them the increasingly differentiated curricula that those judgments require. Second, and more important, in providing these curricula, schools have a responsibility to ensure that access to them is entirely a function of children's own evolving judgments. That is, schools may not forbid children access to any differentiated program of instruction that they offer if such a program is required by children's informed judgments. Of course, access to preparation is no guarantee of success, for in part such success is the responsibility of children themselves, and in part it depends on the nature and extent of the social demand created by others' conceptions of the good.

What emerges from this discussion is a portrait of schooling that begins with common experience of and learning about various conceptions of the good that are live possibilities in children's local contexts, that enables children to find out about and to explore their talents and proclivities that are relevant to making personal judgments about those conceptions, that cultivates the emotional and reasoning abilities to make personal judgments about those conceptions and about the social roles that they entail, that fosters the basic skills necessary to the realization of any of those conceptions, and that develops the political knowledge and skill relevant to constructing and maintaining a social system that permits citizens to pursue cooperatively those various conceptions. As children begin to make judgments among those conceptions and the social roles they imply, schooling then gradually makes available the differential learning opportunities in line with those various judgments. Of course, these judgments are at first tentative and experimental, and the differentiation initially permitted should not be immediately complete and irreversible. Yet as these judgments become gradually more definite, so too should children's school experience become increasingly more disparate.

Unless it seem that this differentiation in schooling is entirely a matter of providing children with distinctive vocational training, it is important to remember that the differentiation of conceptions of the good and thus of children's identities takes place on a variety of different dimensions simultaneously. If, for example, a child begins to affiliate herself with the vocation of software engineering in preference to nursing or the clergy, she may still judge that caring and spirituality are important aspects of her conception of the good even though she does not judge them appropriate as vocations. Thus, an appropriately differentiated schooling must provide for various opportunities to develop these other significant elements of children's personhood at the same time that they also provide suitably diverse vocational opportunities. In other words, the differentiation of schooling does not entail children's segregation into exclusive vocational tracks but instead enables a variety of groupings based on the emerging vocational and nonvocational interests that children develop as their conceptions of the good emerge and evolve.

Even at this point, however, there should be common elements of children's education, particularly concerning citizenship. For, as children develop differential identities, the challenge to them of ascertaining political principles and developing political systems that support those identities becomes increasingly complex, and thus their civic education must prepare them for and involve them in that complexity.

One additional responsibility of central authorities, then, is to maintain a framework for schooling that permits this transition from common to differentiated learning using procedures that guarantee the open access to educational alternatives for which the principle of equal opportunity in education calls. The construction and monitoring of such a framework is a challenging task, especially in light of the threats to equality of opportunity in schooling that our history indicates. On the one hand, this framework must allow children to seek and attain different learning outcomes; therefore, the prescription of and holding schools accountable for uniform student achievement beyond extremely basic skills is inappropriate. On the other hand, these different outcomes are to depend exclusively on the different aspirations that children come to embrace because of their emerging conceptions of the good, their knowledge of themselves, and their understanding of the society in which they live. The only way to ensure this dependence is to regulate and monitor the procedures used in deciding what curricular alternatives are to be offered and in determining the placement of children into them. Specifically, those procedures should make available a range of alternatives that satisfy the aspirations that local children are likely to develop and then provide children with (1) adequate information and counseling about the availability, requirements, and likely consequences of the alternatives; (2) adequate early preparation to enable them to have a reasonable chance of meeting any prerequisite knowledge and skills for the alternatives; (3) adequate opportunities to change alternatives and to incorporate other nonvocational elements within their educational programs; and (4) adequate instruction within the alternatives to give children a reasonable chance of success. In monitoring these procedures, certain outcomes clearly matter because one cannot judge the procedures' adequacy without knowledge of the judgments that children reach, their success in attaining access to the alternatives that they judge most appropriate, and their having a reasonable chance to achieve the goals of the alternatives to which they commit themselves. But the purpose of monitoring these outcomes of the educational system is to ensure *not* that children achieve uniform results or even uniform results up to a particular threshold but rather that the decision making, administrative, and instructional procedures used and the curricula provided do in fact serve the informed and free judgments of the children involved. In other words, the monitoring of these various outcomes of the educational system is useful for determining whether the procedures of that system are adequately conducive and responsive to children's developing personhood.

In summary, the primary responsibilities of central authorities lie, first, with establishing a basic framework for schools that requires and enables schools to pursue the fundamental purposes outlined in the principles of social justice in education. Second, central authorities are to regulate the procedures used by schools in pursuing these purposes to ensure that all children receive an appropriate common education and fair access to differential education that aligns with children's aspirations. Finally, those authorities are to monitor schools and hold them accountable for compliance with these purposes and procedures. Of course, sufficient resources to enable schools to comply in this way are important, and we will consider the mechanisms for providing those resources in the next chapter.

Local Authorities' Responsibilities in School Governance

Local school authorities are to operate within this framework of centrally determined purposes, generally regulated procedures, and accountability. In significant part, the responsibilities of local authorities are to comply with central authorities' directives about these matters, but as we have noted all along, the framework gives localities significant independent authority over the "what" and "how" of schooling to enable children to develop as their own persons and as thoughtful citizens of their communities and the nation as a whole. Localities must operate schools that provide the common and differentiated instruction that we have discussed, but to do so, they must adapt their curricula and their instructional procedures to the cultural backgrounds of their communities and the aspirations that their students develop.

Even the common schooling we have described is not identical in content or method of presentation in different communities or regions. For one thing, the conceptions of the good to which children are to be introduced must be genuine and not simply abstract possibilities for the lives of the children involved. Although there will be a great deal of overlap among them, the particular constellations of such conceptions relevant in individual communities are likely to vary considerably with local cultures, and so, too, is the content of the curriculum for personal liberty likely to vary along those lines. Moreover, the history of conflict and concord among local cultures is likely to be distinctive in different communities, and the instruction needed to enable children to appreciate the relevant conceptions of the good as respectable alternatives for living will necessarily be equally distinctive. If, for example, Jews have been the object of virulent local prejudice, the instruction about their culture and religion will have to be approached differently than in communities where they have not. For another thing, local political arrangements have idiosyncratic histories, emphasizing different problems and strategies for resolving them. Even when the issues are of national scope, local political responses to them can vary dramatically. As a result, locally meaningful citizenship education will need to be adapted in both content and instruction to such histories. Schools in a community with a strong tradition of pacificism, for instance, will need to treat issues of war and peace differently than those in a community with a strong tradition of support for the military if children are to come to have a critical understanding of these complex issues that allows them to take part thoughtfully in the development of an overlapping national consensus about them, with instruction in the pacifist community providing attention to the military issues that are less likely to be accessible in the local community and vice versa.

For differentiated schooling, the content and the approaches to instruction are necessarily diverse. They must, on the one hand, be adapted to children's developing conceptions of the good and their personal judgments about the social roles they will seek in pursuing them. On the other, they need to take into consideration children's prior and independent learning and their aptitude that is relevant to those roles, which can vary not only with children's individual family circumstances but also with the community environment.

Carrying out these local responsibilities for the curriculum and for instructional procedures is a complex intellectual and practical task both for those in charge of local decisions and for those who implement them. For the four principles of social justice, these tasks can be understood as falling into eight related categories, as indicated by the empty cells in the following table. The discussion following the table will necessarily focus on only the first three principles because we have not considered in detail the kinds of education appropriate to the fourth principle, with which we will deal in the next chapter.

Principle of Social Justice in Education	Curriculum Objective	Curriculum Content	Instructional Procedures
Liberty	Developing children's emerging conceptions of the good.		
Democracy	Enabling children to participate in the evolution of an overlapping political consensus.		
Equal Opportunity	Cultivating children's comparative social advantages.		
Economic Growth	Enabling children to understand the role of the economy in the pursuit of their conceptions of the good.		

Although the objectives of the curriculum can be given general formulations in this table, the content and instructional procedures effective for attaining those objectives cannot. For, as we have seen, those matters must be adapted to the cultural conditions and the aspirations of children in each community. At most, then, we can speak only in the most general terms about the processes that local education authorities might use in identifying such content and procedures.

Chapter four outlined one such process for ascertaining the content of education that is consistent with children's developing personal liberty. In brief, this process involves local school officials' determining (1) what liberty-consistent cultures are available in their regions (2) what skills, knowledge, and attitudes local children need to gain access to those cultures (3) what additional attributes these children need to develop to make sound personal judgments about and among those cultures, and (4) what specific learning is necessary for local children to cultivate the attributes required for cultural access and judgment. Once these goals for children's learning have been identified, local officials and educators need to ascertain the classroom and other experiences that will be effective in enabling local children to reach those goals. Of course, this general description makes the process for developing the schools' curriculum and instruction for personal liberty seem much more settled and predictable than it actually can ever be. In practice, this process necessitates, on the one hand, inquiry into local cultures that requires, for instance, sincere communication and negotiation with local members of those cultures and, on the other, experimentation with alternative methods and materials that requires, among other things, teachers' creative thinking and honest self-evaluation as well as their knowledge of the research about effective instruction in other similar communities.

As noted in chapter five, the citizenship curriculum includes multicultural, historical, and philosophical elements, but it too must be adapted to local understandings and conditions. In fact, a similar process of inquiry and experimentation is needed for local authorities to execute their responsibilities for an education in democracy, and it is likely that this process can be completed simultaneously with that undertaken to determine the content and procedures for an education for liberty. Here, too, the history and nature of local cultures need to be analyzed, but in this case the cultures are not limited to those that are liberty consistent. By revealing where consensus will be easiest and most difficult to obtain, a study of the beliefs and practices of all local families' cultures can suggest appropriate political issues to be presented for student inquiry and reflection. These settled and controversial local issues, thus, constitute an important element of the content of the local curriculum for citizenship in addition to other issues that emerge from a consideration of national and international history. In light of this local knowledge, teachers' investigation of and experiments with how instruction can address even the most intransigent local disagreements is possible. It is not the purpose of this instruction, of course, to enforce on children specific, predetermined resolutions of those disagreements, and many of those disagreements, therefore, will remain unresolved as a result of even the most effective instruction in citizenship. Rather, such instruction is to help children understand the grounds of the disagreement and explore the possibilities for its resolution. The inclusion of those disagreements in the citizenship curriculum, children's exploration of those issues, and their thoughtful treatment in teachers' planning for instruction are an important source of children's developing the skills, knowledge, and attitudes that will enable them to participate in the evolution of an overlapping consensus and to understand the personal implications of that consensus for their own conceptions of the good.

As children's tentative and incomplete personal and collective conceptions of the good begin to emerge and as their judgments about these matters begin to take a more settled form, local authorities have a responsibility to provide an education that enables them to pursue the different conceptions that emerge. As we have already noted, the preparation for that pursuit constitutes an education for equal opportunity that will take place in a gradually differentiated curriculum. To plan for and to deliver that curriculum, local school officials need to turn their attention to the specific content and configuration of local children's aspirations to lead fulfilling lives. For these aspirations and the social roles that children judge most suitable in light of their talents and motivations are to provide the framework for that differentiated curriculum rather than school officials' own judgments of, for example, what the best opportunities in the local or national economy may be. Thus, the starting place for planning the differentiated curriculum is a focused investigation of local children themselves. Of course, local officials need to be cautious that children's understanding of their possibilities within their emerging conceptions of the good is not unduly limited. For that purpose, the curriculum should include opportunities to explore the potential implications of their emerging conceptions. Such a curriculum, however, should not take the usual form of career exploration where the focus is limited to the external opportunities that the economy makes available. Instead, it involves children's own exploration of the meaning of those conceptions and their thoughtful consideration of how they may be fulfilled by the opportunities that are likely to be available. Officials should also be cautious in their interpretation of children's aspirations to ensure that their own prejudices about what children from particular ethnic, racial, and gender groups really want or are good at do not limit the curriculum in discriminatory ways. Once they are satisfied, however, that local children's aspirations are reasonably informed and that they genuinely understand local children's preferences, however, local education officials, counselors, and teachers must be equally cautious of substituting for the children's own judgments their adult judgments about what is really best for children.

Once a process has been established for determining and monitoring changes in the various considered aspirations of local children, the planning for and execution of a differentiated curriculum to meet those aspirations can begin in earnest. As noted before, these curricula should not initially produce dramatic differences in children's achievement that would make it impossible for them to redirect their education if a preliminary choice should prove, for whatever reason, untenable. Indeed, the design of the curriculum to provide this gradual differentiation is one of the genuine intellectual and practical challenges that local officials face. Another challenge is for teachers to develop methods of instruction that are capable of optimizing children's chances of success in the alternatives that children find most promising, given their different emerging aspirations and backgrounds. The experience of other, similarly situated communities and the findings of educational research can be of real assistance in meeting these challenges, but ultimately the design and arrangement these curricula and instructional procedures must be a local responsibility, for otherwise the experience of other communities or the preferences of researchers, and not the specific aspirations of local children, are likely to provide the effective rationale for the differentiation, and, as a result, the comparative advantages that children develop are not likely to meaningful and fulfilling to them.

As demanding as these tasks are, probably the most challenging is to create a school atmosphere in which children's diverse aspirations are equally and impartially supported. Equality is provided in part by maintaining the genuinely open access that children have to the various curricula, as is required by the framework established by central authorities, for without children's adequate preparation and without their informed and free judgment, the aspirations of some children will be neglected before they even enter the differentiated curricula. Equality is also provided in part by the adequacy of the methods used for the instruction of students in the various curricula, for without sufficient attention to children's learning needs, their achievement and thus their chances for success will depend more on their circumstances than on their motivation and ability. However, even with open access and adequate instruction in place, children can still be given subtle messages about the worthiness of their choices and their efforts to master the different curricula, messages that can demean some children's cultures and emerging conceptions of the good. Thus, local authorities and teachers must create a school environment in which such messages are neither sent nor delivered. Rather, the explicit and implicit messages must be that each child's ambitions are worthy of respect and that the effort of the school will be to aid each child impartially in the realization of those ambitions. Maintaining such an atmosphere is a crucial responsibility in providing the equal and impartial support that children receive for their disparate ambitions. Accomplishing this task will take many different forms in various communities, depending, for one thing, on the prejudices that are prevalent in them. Here, too, school authorities and teachers need to make themselves consciously aware of the local context and to take action to combat the discouragement that it may present to children's sincere and informed judgments.

Impartiality to diversity is the consistent goal of the principle of equal opportunity in education. In the teeth of the very real challenges that it presents—challenges of curricular design and delivery and of the respect for others with which those tasks must be undertaken, it is all too tempting to misinterpret the meaning of this principle as requiring uniformity in all children's school achievement. But that interpretation neglects both the cultural experience and emerging personhood of many children. Thus, in a society that respects the liberty and citizenship of all, there is no alternative but for local school authorities to undertake these difficult challenges to equalizing children's opportunities in ways that are meaningful to them.

The Role of the Educational Process in Attaining Equality of Opportunity

Some might regard it as surprising or perhaps inappropriate that a chapter on equality of opportunity deals entirely with the regulation of the processes of schooling and with the responsibilities of various political authorities in that regulation. Ever since James Coleman's important work of the 1960s (Coleman 1966), policy makers and scholars have attempted to specify a conception of equal opportunity in education that can be measured exclusively in terms of the outcomes of schooling. As noted here and in chapter three, outcomes *are* relevant to discerning whether schools are delivering equal opportunities, but the specific outcomes of concern must be interpreted in light of the aspirations that children develop as they come to embrace their own conceptions of the good. Some have assumed that all children have similar enough aspirations that the outcomes of their schooling can be measured against common outcome standards in at least some subjects and up to a common threshold. However, we have seen that the proposition that children are in general similar in their aspirations must be erroneous in a society that respects children's personhood and the wide diversity of their cultural backgrounds. Even if there are common skills necessary to children's becoming their own persons and responsible citizens of a democracy—such as literacy, numeracy, and critical thinking—the content on which those skills are to be effectively deployed differs enough with local variations in culture that the appropriate measurement of those skills must be undertaken against standards for the local community rather than for states or the nation as a whole.

Others have assumed that—although there may be much individual and regional variation in aspirations among children of different racial, socioeconomic, and ethnic groups—variations among these groups of children in an entire state or the nation are not significant and that therefore we are justified at least in measuring equality of opportunity against common state or nationwide standards for children's performance among those groups. Indeed, much of the current concern over the achievement gap is based on this assumption that performance differences among those groups allow us to conclude confidently that inequalities in the educational opportunities available to the children in those groups must exist. Now, I have no doubt that real inequality of opportunity is a regrettable fact of American social life given our history and current practice of discrimination and legal and de facto segregation. However, the assumption of similarity in aspirations between the children of these groups must certainly be wrong given the cultural, regional, and even religious differences among these groups. Moreover, this mistaken diagnosis of the nature and consequences of inequalities of opportunity has led us to enforce common standards of performance on these children independently of the judgments that individual children reach about their lives and their ambitions. In any case, even if the considered aspirations of children among these groups implausibly happen to turn out to be similar, the current education policies undertaken to achieve this faulty interpretation of equal opportunity *between* groups of children lead us to neglect the actual variation in children's aspirations *within* those groups. Of course, to a significant extent those differences in children's aspirations may also be, and probably are, the result of unfair discrimination in the society at large and in the schools, but the measurement of children's performance against and the enforcement of common standards in schools do nothing to allow us to detect much less to remedy the effects of such discrimination on either children's aspirations or their efforts to realize them.

In light of these reflections on the conception of equal educational opportunity implicit in the overlapping consensus of Americans about social justice in their schools, we are forced to the conclusion that the only responsible way to enforce the principle of equal opportunity in education is for political authorities, central and local, to attend carefully and conscientiously to some of the processes of schooling—particularly, schools' methods of determining the content of common and differentiated instruction and their procedures for giving children access to the curricula that result from those determinations. There can be, thus, no shortcuts in providing equal educational opportunity or monitoring the extent to which schools meet the requirements of that principle. The imposition

of uniform standards of curriculum content, instructional procedures, or student performance, even if well intended, fails to do so.
Just as problematic is the unreflective importation by localities or
even entire states of the curriculum and instruction in "successful"
schools in other jurisdictions, for such practices, even if they are
genuinely attuned to the aspirations of the children in the jurisdictions in which they are developed, cannot guarantee that the resul-
tant schooling will respond adequately to the emerging personhood
and citizenship of children in one's own locality. Of course, local
education officials and professionals should be prepared to learn
from what works elsewhere, but they need to submit such practices
to careful scrutiny based on a sympathetic and informed understanding of local children and cultures and either to modify those
practices in light of that understanding or to dismiss them out of
hand if they are entirely inappropriate for local children. There is
likewise no effective alternative in achieving equality of educational
opportunity to central authorities holding local schools accountable
for making a sincere effort to achieve that understanding and for
making intelligent judgments about their instructional programs
based on it.

Chapter Seven

Economic Growth and Education: The Financing of Education

The schooling we have described thus far provides an education in and for diversity—diversity of children's conceptions of the good, diversity in children's most personal justifications of the overlapping consensus in their society, and diversity in children's developing abilities to pursue their aspirations. At the same time, several factors in that education draw the society together—mutual respect for others' conceptions of the good and ways of life, the ability and willingness to participate in the evolution of the overlapping consensus, and the impartiality and equality with which children's differential aspirations are treated in the schools. Yet, there is another way in which this diversity in schooling can contribute to diversity and unity in the society, namely, by fostering legitimate economic growth. As we saw in chapter three, economic growth in and of itself cannot be a goal of schools governed by the principles of social justice in education, because that goal enforces the maximization of human happiness or satisfaction as the sole ultimate value of the society, which is inconsistent with the principles' commitment to leave judgments of ultimate value in the hands of citizens themselves. However, suitably constrained, economic growth—that is, growth that permits citizens to enjoy and to protect personal liberty, democracy, and equal opportunity—is a legitimate goal of the schools, for it makes it possible for children when they reach adulthood to realize more fully their personal and collective conceptions of the good. In this way, schooling for legitimate economic growth reinforces the diversity promoted by other elements of children's education by enabling them as adults to realize more completely their differential ambitions. At the same time, such education promotes social unity by helping children understand and appreciate how cooperation between themselves and others in their society facilitates the economic processes whereby this mutual fulfillment in part takes place.

Now, diversity in schooling can itself foster that legitimate kind of economic growth in two ways. First, such diversity can make available and develop the variety of human talents necessary to take advantage of many of the inherently unpredictable opportunities for expanding human productivity. Second, it helps ensure that future citizens have self-defined conceptions of the lives that they find worth living, and those conceptions, in turn, provide a stimulus among citizens to resist certain paths toward increased productivity that are inconsistent with the self-defined fulfillment of citizens' lives. However, such healthy economic growth will not occur without the additional educational preparation of a society's future citizens to understand the connection between their aspirations and the operation of the economy. As we have found that preparation is not simply investment in whatever sort of vocational training that is predicted to maximize growth and then the restriction of educational opportunity to such training, I assume that one prerequisite to that preparation is, first, an education in personal liberty and democracy that will enable children to know and respect themselves and each other, an education described in chapters four and five. For without that education, children will not recognize where to invest their efforts to be productive. I also assume as a prerequisite an education in equal opportunity that develops children's differential and self-determined comparative advantages as described in chapter six. Without this second sort of education, children will not have the capabilities to be productive by their own lights. Against this background, the economic preparation implied by the principles of social justice in education and particularly by the fourth principle involves children's coming to understand the role that their developed capabilities have in helping to realize their own and others' aspirations by means of economic cooperation.

In a market economy, to understand this role is precisely to grasp and to appreciate the operation of the economic system in general and as a whole, for it is, in significant part, by means of that system that the requirements of citizens' various conceptions of the good are expressed and fulfilled. Specific vocational training at most helps children understand only one element of the economic system, namely, the nature and function of a particular profession or at most a few related professions. Even then the emphasis is often on technical matters of performing the activities of the professions rather than on the way in which such professions fit into the lives of practitioners, clients, and the society. An understanding of these other matters, not only for one or a few professions but also for the broader range of genuine possibilities for children's prospective professions and occupations, is what the fourth principle implies. Vocational education can help maturing children understand what a particular occupation requires in the way of skills and activities and even to recognize that those skills and activities have the potential to realize some of their aspirations for a self-defined worthy life. But such education cannot help them to see how membership in that occupation restricts or broadens the opportunities they have to engage in activities outside the labor market by limiting or expanding their available time or by requiring the development of skills and attitudes that may conflict with certain skills and attitudes needed for the other life activities that they also value. Thus, as a result of vocational preparation one is simply not in a position to determine the adaptability of occupations to all the aspects of one's conception of the good. Similarly, children who do not know how and to what extent the activities of a profession assist or interfere with others' realization of their conceptions of the good cannot judge wisely how the occupation fits with their duties as citizens implied by the overlapping consensus in which they participate or with the benevolent aspects of their personal conceptions of the good. As a final example, if children do not know about the economic returns to an occupation, they cannot fully judge whether membership in that occupation will help them realize their self- or other-directed aspirations outside the occupational sphere. These reflections suggest the need for children to achieve a wider understanding of how the economy coordinates the activities of individuals and groups and whether or how it enables them to realize their aspirations.[1] The achievement of such an understanding further suggests that the broader education required to serve this principle can be provided most adequately by instruction in economic history and theory. Such learning can provide the broad understanding of potential occupations and their social and personal consequences required for children to make sound decisions not only about their own futures but also about consequences for the overlapping consensus in their society of the economic policies that their society might pursue. Thus, the principle of economic growth in education, the seemingly most vocational element of the conception, implies profound intellectual content, not just narrow job preparation.

Of course, economic education as it is currently often practiced in American schools takes a highly ideological form that promotes an acceptance of the dominance of the marketplace over individual and collective ambitions. Economic history and theory, however, do not necessarily support such a bias. After all, they include the study of a whole host of market failures and the governmental mechanisms that are available for the society to correct for such problems. Moreover, these subjects, broadly understood, do not involve indoctrination into an unwavering normative commitment to utilitarianism, as the wider normative bases of economic theories now available demonstrate, such as that of Amartya Sen (2002), for example. As already suggested, a broad education in economic history and theory can provide an important complement to the education for personhood and citizenship that we have described by helping children to understand the possible personal, social, and economic consequences of their putative judgments about conceptions of the good and the issues of the collective control of the economy that will face them as citizens. Further, economic education of this kind can contribute importantly to the judgments that children make as they develop their comparative advantages by enabling them to understand the economic value of various aspects of their potential and the economic costs of its development. In fact, elements of such an education were implicit in chapter six's discussion of the schools' responsibilities for equal opportunity, where helping children to have informed differential aspirations included, in effect and among other things, an understanding of how their talents and motivations interact with labor markets.

When we formulated the four principles of social justice in education in chapters two and three, we noted that the principles work together to deliver social justice. Meeting the requirements of one of these principles in the absence of simultaneous attention to the requirements of the others produces an imbalance in children's education that amounts to injustice. We have reinforced that conclusion in our discussion of each of the principles individually. For example, equalizing children's opportunities but neglecting their rights to develop as their own persons and as participants in the overlapping consensus can have the consequence of demanding uniform or otherwise externally specified school achievement that ignores the personal and political meaning of such achievement. Similarly, enabling children to participate in an overlapping consensus but depriving them of the opportunity to understand the economic consequences of their deliberations encourages a kind of utopian thinking that neglects a crucial social dimension of political decision making. By the same token, our discussion of the implementation of the principles in this and the last three chapters has revealed the extent of the practical interdependence of those principles. As we have seen, the school practices implied by each of the principles reinforce and capitalize on those required by the others. This result is especially notable in legitimate economic education, where we have found that such education effectively harmonizes with and, in a real sense, completes children's education for personhood, citizenship, and comparative advantage. This mutual reinforcement of the practical implications of the four principles is an important virtue of this account of public schooling in that it provides for a coherent understanding of not just our theoretical responsibilities for education but also the practical activities that those in positions of authority must undertake to meet those responsibilities.

Now, it might seem that this economic education is a place where uniform national standards for the curriculum and for children's mastery of it might have a role. However, this discussion of the way that economic education integrates seamlessly into the sorts of education we have found to be consistent with the other principles of justice in education contradicts that conclusion. Thus, local authorities' knowledge of their children's emerging conceptions of the good and the judgments they are to make about their comparative social advantages is crucial to determining the elements of history and economic theory that are most relevant to maturing children's lives. Similarly, local authorities' knowledge of their children's confrontation with the current overlapping political consensus, of their emerging assessments and criticisms of that consensus, and of the way that their developing conceptions of the good shape their efforts to ascertain and to justify a more adequate account of political values is equally crucial to determining the economic content that is most politically meaningful to them. Thus, as was the case with the other principles of social justice in education, it is not possible to specify in more than an abstract and general way the goal of the economic element of schooling, that is, to enable children to understand the role of the economy in the pursuit of their and others' conceptions of the good by means of helping them understand the elements of economic history and theory that are relevant to their evolving personal and political judgments. Therefore, the delineation of the precise content of the curriculum and the methods for teaching it must depend on local authorities' investigation of the emerging economic interests of the children for whom they are responsible coupled with their intelligent judgment about how to make economic history and theory serve the aspirations that their children are developing. Of course, as we have seen with education that meets other principles of social justice in education, central authorities should hold these local authorities accountable for pursuing the legitimate economic purposes of education, for conducting their investigation of the economic dimension of local children's aspirations honestly and sensitively, and for applying their professional judgment about the economic curriculum and effective economic instruction in ways that are true to what is known about their children's learning. But in this case, too, the ultimate responsibility falls on local authorities and professionals to determine within these general guidelines the content and procedures of the economic education that is required to enhance the personal and political lives toward which local children themselves are developing.

To clarify just how local authorities might accomplish these tasks of developing the economically relevant curriculum, it would be helpful to consider a suggestion that Harry Brighouse (2006) has made. He argues that educating children for participation in the economy can contribute to children's well-being not only in that it can provide the income that is needed to help realize their aspirations but also in that it enables children to engage in the kind of work that itself realizes those aspirations and in that it provides children with a sense of self-reliance that he believes to be necessary for the satisfaction that they derive from their lives. However, he, too, has proposed that the role of economic growth in developing the school curriculum should be limited in children's own interests. Specifically, he notes that social science has observed that such growth beyond a certain threshold proves to be of little value in improving what he calls citizens' subjective well-being, that is, the sense of satisfaction that they derive from their activities and their lives in general. Thus, in advanced economies that have already reached that threshold, he argues, educating children to contribute to the growth of the economy should be balanced with educating them for activities that they find satisfying outside the world of work, that is, including a life of productive leisure.

Assuming that the social science research on which Brighouse relies is trustworthy, we now should consider whether local education authorities may use such research in the design of the general school curriculum and particularly of the economic dimensions of that curriculum. Now, much of what Brighouse asserts, and particularly his rejection of the economic theory of human capital, aligns in a general way with the conclusions we have reached. Brighouse's conclusions, however, derive from what he takes to be a kind of technical inadequacy in that economic theory and not the inadequacy of the underlying normative justification of the theory itself. That is, Brighouse objects to the assumption of human capital theory that economic production always provides a reliable index of human life satisfaction, particularly in advanced economies. In such economies, Brighouse believes, noneconomic activity becomes an increasingly important source of satisfaction for which human capital theory simply cannot account. Human capital theory fails, on Brighouse's account, because it cannot satisfactorily explain all the circumstances on which the subjective well-being of humans depends and *not* because it makes such a sense of well-being an ultimate political value. Children's education may, therefore, be engineered in the interests of increasing subjective well-being, as Brighouse predicts will happen if school authorities enforce a healthy balance between work and leisure. Of course, that does not necessarily mean that Brighouse assumes that subjective well-being is the exclusive value in education, as he clearly does not, but it does disclose that he does think that education officials are justified in designing schools to realize such values in children's lives. That conclusion clearly conflicts with the principles of social justice in education that have been developed here because a socially just education is to enable and to allow children to reach their own judgments about the ultimate values of human lives as they develop and embrace their own personal and political conceptions of the good.[2] Thus, judgments about whether and the extent to which subjective well-being is an important value are to be made by maturing children and not by political authorities or education officials.

As a result, local officials must be cautious about the explicit or implicit assumptions they use in developing the economic or noneconomic elements of local schools' curriculum and instruction. In particular, they must be alert to the normative implications of any social science they employ in making these decisions. Of course, any particular social science findings they might be tempted to use need not inherently assert that certain values are of supreme political interest, but such findings may well do so. Even if they do not, however, in basing their decisions on such social science local officials may unwittingly be fostering norms of dubious political value to the overlapping consensus about education in our society. Thus, Brighouse's analysis is an attractive but dangerous example of the kind of reasoning that local officials should avoid. Instead of seeking to engineer children's values in the belief that they and the social science they consult have insights that children do not and cannot have, such officials should attend carefully to the non-public values that children are actually developing in the contexts of their local cultures and communities, values that are in addition to the public values that are included in the overlapping consensus as represented by the four principles of justice in education. The school curriculum, then, should be designed to enable children, first, to make their own judgments about such values and, second, to pursue the values in and by means of their education that children judge to be important.

The Inadequacy of the Standard Economic Approach to Financing Schools

One important task that we noted but did not undertake in the last chapter is to develop an approach to providing and distributing the resources required by the education we have described. Now that the entire account of education for social justice is available, we can turn to that task. Clearly, economic theory might be relevant to determining and distributing educational resources, but we need to be thoughtful and cautious about our application of that theory because such an education seeks to constrain the types of economic growth that a society pursues in the interests of the diverse aspirations that this education makes possible among the society's children.

The standard economic analysis of the financing of public social services suggests theoretical answers to the critical questions of how much should be invested in these services, to whom the funds should be distributed, what means should be used to allocate that investment, and who should pay for it.[3] On that account, the optimal public investment in those services is the amount that will return the greatest long-term net economic benefit to the society. In education, as with other social services, both private and public expenditure can improve people's economic productivity, but individuals have an incentive to make private expenditures only to the extent that they themselves are likely thereby to receive future individual income (or what economists call internal benefits) equal to or greater than their initial investment. However, at least four factors suggest that private expenditures on education will be insufficient to maximize the economic return to the society as a whole.

First, individuals may be ignorant of the internal returns to education. Here, in calculating the amount they are willing to invest in education, individuals may simply not be cognizant of many of the individual returns to their own education, and underinvestment will result. Second, the individual returns to education, even if they are known to individuals and generally predictable, are uncertain. Thus, while a high school graduate will earn on average a particular amount more than someone who has not graduated from high school, about half of high school graduates will earn more than that amount and half will earn less, and an individual cannot be sure of his exact place in the range of eventual post-high-school incomes. The society would be best off if individuals would use the average return in calculating the amount that they will expend on a high school education, but some individuals may use a lower figure, perhaps even the minimum return to high school education, in their calculations. Therefore, if enough people choose to use lower than average estimates (as they would be reasonable in doing if the median return is lower than the average return), uncertainty, too, can create underinvestment in education. Third, many of the economic returns to education benefit people other than those who become educated (or in economists' terms, some of the returns to education are external). For example, a highly educated medical researcher may use her education to develop a cure for a disease that will benefit untold future generations of disease victims, but her current salary is not likely to reflect the huge value that all those future beneficiaries receive from her work. But in deciding how much to invest in their medical education, individuals are likely to use in their calculations only their expected future income as medical researchers and not the expected total social benefits. Thus, the external benefits of education can also create underinvestment. Fourth, the current distribution of family wealth does not necessarily match the potential of individuals to generate future education-dependent returns. For instance, a young man with the potential to revolutionize the field of materials engineering may come from a poor family, which is likely to mean that he cannot afford to attend the engineering school that would develop that potential. In determining what they will expend on education, individuals are likely to use what they can afford rather than the benefits that they could create in their calculations. Thus, here too, the current distribution of wealth can create underinvestment in education.[4]

For reasons such as these, economic analysts conclude that public expenditures on education are required to maximize its net economic benefits. And the total public expenditure required is equal to the precise amount of underinvestment in education that would occur if all education decisions were made entirely by individuals on their own. In theory, these public expenditures should be used to change the incentives that individuals have to underinvest in their education so that they actually obtain the education that their potential to create future economic returns implies. Specifically, these public expenditures should be directed to those whose education decisions would be limited by a lack of awareness or the uncertainty about the internal returns, those who have the potential to create significant external returns, and those whose access to wealth is limited by their family circumstances. In other words, public expenditure for education should be directed to the uninformed, the risk averse, the public-service oriented, and the poor. It is likely that everyone in a society is affected by these factors to some extent, especially when they are very young and lack the ability or motivation to make accurate predictions about the internal returns to education. Thus, there is, on this account, justification for general public expenditures to support the education of the young. But as individuals grow older and become as result of this early education less ignorant of the internal returns of their education, public expenditures should be targeted to those whose decisions are most affected by risk aversion, external returns, and poverty. Therefore, there is on this account justification for the public to subsidize the more advanced education of, for example, the children of families with limited means and those who choose to enter public service professions.

These public expenditures can be used to compel individuals to obtain education, to subsidize the education that they receive, or both. The need for compulsion is greatest in the case of young children because they lack the ability and motivation to consider even the internal returns to education. But the case for compulsion weakens as such ignorance is reduced because these individuals come to have the ability to make rational decisions in light of the knowledge of their own potential and its economic consequences for themselves.

In fact, it can be argued that older individuals are in the best position to understand their own potential and motivations. Thus, the economically preferred use of public expenditures gradually changes from compulsion coupled with subsidy for improving individuals' ability and motivation to respond to the economic benefits of education to the provision of incentives and subsidies for individuals to take advantage of the education that can optimize the internal and external benefits in the society at large.

On this account, the costs of public services should be allocated to those who receive the economic returns of those services because the net benefits can be maximized only if those who receive the benefits also bear the costs of creating them. Otherwise, the costs incurred may exceed the benefits received. In education, the determination of who should pay for these public expenditures depends on whether the returns to the education provided are internal or external, that is, on whether the returns will accrue to the educated individuals themselves or to others. Compulsory education for the young has both kinds of returns because it provides the foundation for subsequent education. However, the returns to the public expenditures to correct for the uncertainty of the returns to education and for the distribution of wealth are entirely internal because these factors are defined as problems in individuals' ability or willingness to invest in education up to the full amount of its future internal returns. Also by definition, the returns for the final category of public expenditures are entirely external. Thus, the returns to education to correct these four problems are a mixture of internal and external benefits, and the costs should be borne both by those who are educated and others. But even costs associated with the internal benefits of education cannot be allocated immediately to those who are becoming educated, for otherwise the problems of ignorance, risk aversion, and poverty would not be corrected. Thus, the public needs, on the one hand, to bear much of the costs of such education at the time that the education is provided and, on the other, to collect those costs from the educated individuals in the future. Income and wealth taxes are plausible vehicles for collecting these costs because incomes and wealth vary roughly with the internal education returns that one experiences. However, the external benefits that one receives from others' education do not vary to a significant extent with one's income and wealth, and for these a uniform per capita tax is probably the most appropriate way of assessing costs.

The standard economic reasoning about the public financing of education, although highly abstract, is attractive in that it appeals consistently to a normative criterion to answer questions about the optimal public expenditure, its distribution, the means by which that distribution can be effected, and the allocation of costs—namely, the maximization of the net economic returns to education—and it generates systematic if very general answers to those questions. Of course, this generality implies that a great deal of hard work remains before this reasoning can produce a practical plan for financing education—estimating the actual amount of the expected net benefits of education, designing effective procedures for distributing the education that will provide those benefits, and so on.

However, this reasoning is an inappropriate basis for financing socially just education because it violates each of the four principles. The standard economic analysis of education funding violates personal liberty and democracy as we have conceived them because it enforces an ultimate value that members of society in their roles as independent persons and as citizens may not judge to be worth pursuing, at least as the sole and highest aim of their lives. It also violates our conception of equal opportunity because the educational opportunities it makes available are inconsistent with many people's judgments of what their personal and collective conceptions require. Finally, it even violates our conception of politically legitimate economic growth because it may enforce the development and deployment of abilities that are inconsistent with the realization of many citizens' conceptions of their own and the society's good. It is clear, therefore, that the four principles require a different approach to school funding.

Financing Socially Just Schools

One value of our review of the standard economic analysis of education finance is that it reminds us of the general questions that any such financing system should answer:

- What purposes should the system achieve?
- How much funding should the system provide?
- To whom should the funding be distributed and in what amounts?
- Who should bear the costs of funding education?

These questions provide a useful framework for the analysis of the financing system required by the principles of social justice in education.

The answer to the first question is straightforward because the four principles articulate the purposes of public education and thus of the financing system in a socially just society. The purposes of financing education are thus four—to enable children to become their own persons, to participate in the evolving overlapping consensus, to develop their comparative advantages, and to understand the role that economically valued capacities have in formulating and pursuing their personal and social conceptions of the good. It should be noted that none of these purposes posits an ultimate value that schools are to enforce on children. It might be tempting to conclude that personal autonomy is an ultimate value implied by these purposes, given their emphasis on personal and political liberty. But, as we have found, such a conclusion would mischaracterize the argument that has been provided for the principles and the meaning of the principles themselves. The justification for the principles started from what members of this society typically believe that schools are for, not from an effort to establish the rational superiority of any particular value or set of values. In John Rawls's terms, it is a political and not a metaphysical justification. As that justification developed, it was necessary to clarify, limit, and specify the meanings of those beliefs because, without such restrictions and stipulations, they often were in conflict with one another. That process of rendering these beliefs consistent with each other, however, did not appeal to or attempt to identify ultimate values but rather to elucidate what seems to be implicit in citizens' expectations of public education. The principles that emerged from this process, then, were an articulation of the current overlapping consensus about schooling. Moreover, as we have seen in previous chapters, the liberty implied by the principles of this consensus does not amount to a radical independence of children from the context in which they develop, as autonomy taken to be an ultimate value seems to require. Rather, that liberty aims at personal and political freedom within the context of the cultures that children encounter and perhaps embrace as they develop into their own persons and into citizens of their society.

The answer to the question about how much funding should be provided is complicated. As already noted, socially just education must provide for the development of five related aspects of children's knowledge and ability—(1) knowledge of the cultures relevant to children's lives and of their own talents and proclivities and the abilities to make reasonable personal judgments about those matters in developing a personal conception of the good, (2) knowledge about the history and current state of the overlapping consensus and the abilities to participate in its further evolution, (3) knowledge of the basic skills relevant to the pursuit of any conception of the good and the abilities to apply them to the realization of children's own aspirations, (4) knowledge of children's comparative advantages and the abilities to use these advantages to pursue their emerging personal and collective conceptions of the good, and (5) knowledge of the economic dimensions of children's aspirations and the abilities to use that knowledge in fulfilling their emerging conceptions.

At this point, two general observations about the education needed to provide this knowledge and to develop these abilities are relevant to the task of determining the funding of that education. First, the education to achieve these goals inherently varies with a great many factors. Thus, for example, the cultures about which children are to become knowledgeable are likely to be common in particular geographic areas, but they may differ dramatically from one jurisdiction to another. Similarly, the basic skills and much of the knowledge of political and economic systems to be learned are likely to be similar for all children, but adequate instruction in those subjects will have to accommodate differences in the application of that learning because children's cultures and emerging conceptions of the good differ. Moreover, even in those cases where the knowledge and abilities to be gained are identical because children hold the same aspirations, the instruction necessary to achieve that learning is likely to differ with the advantages and disadvantages that children face as a result of their growing up in particular families and communities. Thus, despite the general descriptions that follow from the principles of social justice, there is such built-in variation in the outcomes and processes of a socially just education that it is difficult if not impossible to identify a common cost or even an average cost for the education that will achieve the purposes specified by the principles. Second, this problem is compounded by the fact that the achievement of these purposes is in all cases a matter of degree, not a matter of meeting a fixed goal for or a minimum threshold of learning. Thus, for example, children can become more or less knowledgeable of locally relevant cultures or more or less able to pursue their conceptions of the good. We can agree that having no knowledge of the various locally relevant cultures is inconsistent with children's becoming their own persons but are unable to determine with certainty and in principle just how much knowledge of that kind is sufficient to guarantee that all children actually become their own persons. Or we can agree that every child needs a fair start in developing the differential abilities required by his or her conception of the good, but we cannot determine in principle just how extensive the public responsibility for developing those abilities should be. Since the extensiveness of this education cannot be settled by the principles alone, neither can the costs of that education.

Because of these two factors—that education legitimately varies with a whole host of circumstances and that the extent of education required to meet these purposes is indeterminate—it becomes impossible to establish in principle and with any certainty the total cost of such education. This result might tempt us to conclude that an answer to the question about the total funding for education is impossible, but that conclusion would be hasty. What these factors really suggest is that the answer to this question is not exclusively a matter of principle but, to a significant extent, of political judgment (cf. Gutmann 1999). In other words, above some constitutional minimum perhaps, representative political authorities are to determine the total funding for schooling, bearing in mind the purposes of schooling that are articulated in the four principles of social justice in education. These two factors, moreover, imply that a constitutional minimum cannot be reliably specified beyond the requirement that governments establish and maintain a system of compulsory schools that pursue those purposes. As a result, the total funding for compulsory schooling should be in the hands of appropriately constituted political authorities guided by the overlapping consensus on education.

As the content and procedures of schooling vary legitimately from one community to the next, it is tempting to conclude that local authorities should determine the funding for schools in their jurisdiction, but this conclusion is also hasty. After all, the basic purposes of education must not be allowed to vary with local tastes for or preferences about schooling, and, therefore, the responsibility to determine total funding falls to central authorities, who are most clearly subject to the societywide consensus on schooling. In other words, representative central authorities should determine the total funding that is necessary to meet the purposes of public schooling, and local authorities should determine how the funding available to them is allocated to curricular and instructional programs while they are held accountable by central authorities for meeting those purposes as effectively as that funding allows.

The next question is how that centrally determined funding should be distributed to localities. Nothing in the argument thus far implies that the funds should be distributed by central authorities in identical amounts for each student in local schools to achieve what is called horizontal equity in the school funding literature. Indeed, such equality of school resources between schools or school districts is only in the rarest circumstances an acceptable index of the financing system's contribution to the purposes of socially just schooling, for both the material and cultural conditions under which children grow up are likely to affect schools' ability to achieve these purposes, and such conditions usually vary by school and geographic region. The distribution should, therefore, seek to provide schools with the funds they need to achieve the requirements of social justice. Thus, the basic distribution formula should provide more funds to schools with higher expected educational costs to meet the basic purposes of education and less to those with lower costs. Again using the language of the school funding literature, the formula should aim at a special form of what is called vertical equity, that is, the unequal distribution of funds that provides schools with an equal chance of meeting the social justice purposes of education regardless of the conditions that children, families, and communities face in doing so.

Some factors that generate more educational cost are familiar elements of current state distribution formulas, such as the incidence of child disability and family poverty. Others may be less familiar, such as the level of educational attainment in the community that may necessitate additional educational expenditures or the sparseness of population of some areas that may require smaller than average class sizes and the additional expenses they involve. Some, to my knowledge, are not included in current formulas at all. For example, the more insular some families' cultures are or the more isolated particular cultures are within the society, the more difficult it will be for schools to enable children from such families and cultures to develop a respectful regard for at least some other cultures in their society and to teach other children to respect those cultures. Such schools are likely to require more resources so that the children in those regions can become their own persons, respect others' personal liberties, develop as citizens who are willing to teach about their own communities' political endeavors and to learn from those of other communities, and identify and develop their comparative advantages in the context of the wider society's economy. In any case, the responsibility of central authorities is to identify as completely as they can the variables that affect schools' costs of providing a socially just education and to distribute funds in accordance with those variables. Whatever the precise details of the distribution formula turn out to be, it is clear that a common feature of current school funding is simply inappropriate, that is, the allocation of more funding to communities with greater real property wealth. At best, property wealth may be irrelevant to the cost of achieving the sort of vertical equity we have been discussing; at worst, that cost may be inversely related to such wealth.

Moreover, not every allocation scheme that aims at some version of vertical equity is justified by these principles. For example, another recent standard, adequacy, is also deficient from a social justice perspective. Adequacy is often defined as requiring the state to distribute sufficient funding to enable all children to achieve at least the minimum level of academic performance established by state standardized testing requirements. An adequacy-based formula would distribute more funding to schools with students who have difficulty in meeting that minimum and thus also would provide a form of vertical equity. As we have already found, however, the goal of attaining a minimum uniform level of student achievement according to state standards is inconsistent with social justice in a number of ways, most notably because it likely enforces the political majority's conception of the good on all children and, therefore, violates the principle of personal liberty in education and because it completely neglects the education necessary for children to develop their differential comparative advantages and, therefore, violates the principle of equal opportunity in education. Thus, the sort of vertical equity implied by the standard of adequacy so defined is inconsistent with the vertical equity required by social justice.

The final question about school funding concerns who should pay the costs of socially just education. As the provision, purposes, and nature of the schooling described here follow from the overlapping consensus about education, it is appropriate that its cost be borne by the entire body of citizens who participate in that consensus. The argument for this education does not parse its benefits into those that redound to private individuals and to the society more generally. Rather, the obligation of children and their parents to participate in that education and the obligation of citizens more generally to provide such an education are not conditional on the specific economic returns of that education. In other words, it does not follow that the obligation to participate is reduced if a child or her parents can demonstrate that the predicted individual returns are limited. Nor does it follow that the obligation to provide and pay for such education is reduced if its predicted combined social and individual returns are low. As a result, the costs of this education are allocated to the members of society collectively rather than to the individual members of the society in proportion to the benefits they receive.

Of course, in a largely market economy, it will be necessary to raise the revenue to meet those costs from private citizens through taxation because, by the definition of a market economy, the government does not maintain sufficient economically productive resources to cover those costs on its own. Inevitably, such taxation will need to vary with citizens' ability to pay, and there is a variety of vehicles for collecting the required revenue—income, consumption, and real property taxes, for example, all vary with some measure of citizens' ability to pay. However, the precise configuration of such taxes cannot be determined by the principles of justice in education themselves. Here, as was the case with determining the total expenditures for education, the forms of taxation to be used are largely a matter of political judgment rather than principle, and because the total costs are appropriately determined by central authorities, so must the precise configuration and rate of taxes to cover those costs be centrally decided.

In summary, the purposes and the general scheme for the proper distribution of school funding are logically derived from the principles of justice in education even though the details of the education required to meet those purposes and the precise distribution formula cannot be derived from the principles alone but must take into account the circumstances that exist in various communities. The decisions about the total amount of funding for schools, the specific variables of the formula for distributing that funding to local schools, and the allocation of school costs to taxpayers are to be made by democratically constituted central political authorities who bear in mind as they make their decisions the purposes of schooling and the varying local circumstances that affect the achievement of those purposes. As we saw in chapter six, control over the purposes, content, and procedures of socially just education is appropriately divided between central and local authorities, with significant responsibilities for determining educational content and process falling to localities that are held accountable to central authorities for the purposes and effects of their decisions about those matters. In school funding, however, the balance of responsibility lies largely with central authorities, with local authorities being expected to make wise use of the resources made available to them to accomplish the purposes included in a socially just education. Moreover, our analysis of those responsibilities is very different from the standard economic analysis because the purposes and allocation of education turn out to be dramatically different. Economic thinking has an important place in socially just education in that the diversity that results from that education is an important source of legitimate economic growth and in that economic history and theory are part of the content of that education. But standard economic theory is an inadequate guide to the overall substance and procedures of education and, thus, to the funding and its distribution appropriate to providing to children the education that social justice requires.

CHAPTER EIGHT
SOCIALLY JUST EDUCATION IN A REPRESENTATIVE AMERICAN COMMUNITY

Now that the account of the principles of social justice in education and their implications for many of the general policies and practices in schools is complete, we are in a position to determine how and to what extent social justice has a role in resolving some of the controversies that arise over contemporary schools and school systems. For that purpose, I return to the situation in Jamesville, not because the problems in that hypothetical case provide anything like an exhaustive list of the possibilities for disagreement over education in this country but because the controversies briefly described in chapter one are typical of those that arise in American schools these days and because, I hope, these problems and the ways in which the principles developed here approach them are illuminating about the meaning and reach of social justice considerations about education in the United States.

To recapitulate, the specific controversies over Carrington Middle School and more generally about all Jamesville's schools arise from a convergence of the increased population of Latino students in Jamesville's schools with the new accountability requirements of the state that follow from the federal No Child Left Behind Act. On the one hand, Carrington's annual accountability reports show its failure to meet the state's definition of Adequate Yearly Progress (AYP) for three consecutive years, which means that the mandatory school improvement plan instituted after its second year of failing to meet these requirements has not yet worked sufficiently well to correct the deficiency in test scores. The school will now have to begin using some of its Title I funds to offer its students free tutoring, and it faces the prospect, if another consecutive year of failure to achieve AYP occurs, of having to replace much of its teaching and administrative staff, among other possibilities (Hess and Petrilli 2007). On the other hand, the state accountability reports also show that Carrington has not achieved AYP because an insufficient proportion of its students with limited English proficiency, largely its increasing number of Latino students, have not scored well enough on the state's standardized tests in English and mathematics. Moreover, their test scores in these subjects have not improved at the rate prescribed by the state. As we have seen, administrators, teachers, parents, and other community members in Jamesville have raised a host of issues about this situation, issues about the fairness of the accountability system, the availability of resources to deal with the problem, the academic focus of the schools, and the adequacy of the instruction provided to Latino students. Moreover, the positions taken on these issues point to different, and sometimes opposing, courses of action, for example, modifying or even repealing state accountability policy and reallocating or increasing resources in the schools.

At first glance, two principles of social justice in schooling appear to be significantly implicated in the controversies in Jamesville—the principles of liberty and of equal opportunity. A concern about liberty seems to be raised by the accountability system's imposition of uniform student performance requirements as measured by standardized tests, regardless of what parents want for their children or what children want for themselves. A concern about equal opportunity seems to be raised by the differences in test performance between Latino and Anglo students. We should be cautious and precise, however, in making these applications of the principles.

Although social justice in education is deeply respectful of student diversity in achievement, at times each of the principles, at least in the context of the realities of American society, calls for a degree of uniformity in student performance. For example, children's personal liberty in this context implies that they need the basic literacy and numeracy that give them genuine access to the cultures that provide authentic possibilities for their conceptions of the good. Similarly, equality of opportunity also requires these skills as a basis for children's pursuing the differential curricula to which their emerging conceptions of the good may lead. Beyond this, the principles of personal liberty and democracy suggest that children need to develop skills of perception, reasoning, and argumentation that enable them to reach sound judgments about, on the one hand, the alternatives for their conceptions of the good and, on the other, the current content of the overlapping political consensus and possible modifications in it. Finally, the principle of economic growth implies that children need a sufficient command of basic economic history and theory to help them determine whether various occupations and economic arrangements are consistent with the requirements of their personal and collective conceptions of the good.

In light of these commonalities in socially just education, the mere fact that an educational policy or practice aims at a degree of uniformity in children's achievement does not necessarily mean that it is unjust. The important questions about such uniformity are really two: First, is the uniformity at which the policy or practice aims consistent with the commonalities specified by the principles? Second, are the particular curricular and instructional mechanisms used to achieve that uniformity consistent with the eventual diversity of aspirations and achievement for which a just education must also provide? Thus, a requirement of uniformity can go wrong in two ways. It can impose the wrong kind of commonality on children's education, and it can impose the right kind of commonality in a way that discourages children from becoming their own persons or thoughtful citizens and from pursing the differential possibilities for their lives that they judge worthy according to their emerging personal and collective conceptions of the good.

In assessing the state-imposed uniformity in student achievement in Jamesville, then, we need to consider, first, whether those requirements are a legitimate part of an education that intends to liberate and enfranchise children or whether those requirements intend, instead, to constrict unduly their possibilities as persons and as citizens. And here the analysis becomes complicated. One possibility that must be considered, for instance, is that localities left to their own devices are likely to discriminate against some children in ways that deprive them of the basic learning on which their personhood and citizenship depend. Another possibility that must be considered is whether such requirements effectively discourage some children from taking their own family cultures or those of other residents of their community seriously in developing their personal and collective conceptions of the good. Either possibility represents a genuine injustice. However, these possibilities cannot be determined in the abstract; rather, they must be assessed by considering the specific facts of the situation in Jamesville.

From what we know so far about Jamesville, we have evidence that both of these possibilities may be in operation there. As we argued in chapter four, current state academic performance standards inevitably reflect dominant cultures, whether they are intended to do so. Thus, even though these standards focus on literacy and numeracy, they are typically not developed in a way that limits their content to what is necessary for children to gain access to the various cultures that might be resources in developing their conceptions of the good, nor are they adapted to the differences in configurations of culture that are relevant in various geographic regions and localities within the state. In this case, the state's standards were likely formulated to reflect the political majority's beliefs about what is important for the lives they believe are worthy, not necessarily considering what kind or level of literacy or what written resources, for example, Jamesville's children need for them to understand the various traditions of their own and other local families so that they can appreciate those traditions as possibilities for their own lives, make reasonable personal judgments among those traditions, and come to respect these traditions even if they do not adopt them as sources of the conceptions of their own good. As a result, these state standards subtly encourage children to view the meaningful possibilities for their lives as restricted to those acceptable to the political majority in the state as a whole, thereby neglecting other meaningful possibilities that may be available to children in their local communities. The narrowing of the Jamesville community's conversation about their schools to local children's performance on state standardized tests and the instructional arrangements needed to improve it suggests that any citizen's concerns about children's access to cultural alternatives not implicit in the state standards are simply off the agenda. Moreover, the possibilities for Jamesville's children's developing conceptions of the good are concomitantly restricted, and their knowledge and appreciation of those who practice local cultures that fall outside the official state definition of what is acceptable is also limited, even if those local cultures are perfectly consistent with others' personal and political liberty.

At the same time, the reaction of Jamesville's Anglo parents to the schools' unsatisfactory test scores is evidence that discrimination is also present in that community. Those parents interpret the schools' efforts to adjust their instructional programs to the realities and needs of the new Latino population as a diminution in the schools' commitment to the traditional academic curriculum, with the consequence that their own children, they believe, are no longer receiving the kind and quality of education that will prepare them adequately for the ways of life that those parents prefer for them. However, the culture that this system of accountability enforces on all children is, in fact, the Anglo parents' own, dominant culture. In effect, these parents seem to want to keep what they think of as the real advantages of access to that culture for their own children alone by denying to Latino children the modified instruction that may give them access to those advantages. Nor apparently do they care that those adjustments in instructional procedures might, although unintentionally and to a limited extent, enable their own children to become aware of, to appreciate, and perhaps even to adopt possibilities for their lives that Latino cultures make available. In short, they simply do not appreciate the potential that access to Latino cultures has for broadening the possibilities for Anglo children's personhood or strengthening their citizenship. In this way, the reaction of many Anglo parents in Jamesville reveals the presence of subtly discriminatory attitudes in that community, attitudes that lead those parents to reject the schools' instructional adaptations to Latino students even though those adaptations explicitly enforce the dominant culture.

Ironically, while the state standards themselves reflect an enforcement of the dominant culture on all children, the state accountability mechanism adopted as a result of NCLB directs the schools' attention to at least some of the possible needs of children who are not proficient in English. Admittedly, those needs are defined in terms of what it would take to allow Latino students to gain access to the dominant language and culture, but at least the accountability mechanism ensures that local school officials must become knowledgeable about and sensitive to the Latino community in ways that they otherwise might not. In part, this irony may explain the quiet but concerned response of Jamesville's Latino community to the situation in their schools. First, they see the controversy as revealing discriminatory attitudes in that Latino children and, indirectly, their families have been blamed for the problem. Second, they also see that the state accountability mechanism requires local officials to pay special attention to their children and to change the schools' instructional programs in ways that will enable those children to succeed in meeting the testing requirements. But, finally, they also see, at least intuitively, that that particular definition of success does not necessarily encompass all, or even much, of what parents in the Latino community seek for their children. For these reasons, members of the Latino community have an ambivalent response to the state standards and the steps that schools in Jamesville have taken to meet them. They may appreciate schools' additional efforts on behalf of Latino children, but they cannot wholeheartedly or unanimously endorse the purposes for which those efforts are undertaken. They may worry about the sincerity of those efforts considering the discrimination that they and their children are experiencing. In effect, the Latino community faces a forced choice between two unsatisfactory alternatives—school accountability for results that do not fully represent their members' aspirations for Latino children or a lack of accountability that leaves localities free to neglect Latino children's interests altogether if they so choose.

One might conclude from this analysis that the principles of social justice in education are, in practice, hopelessly inconsistent with one another. For it seems that the attempt to provide children with equal educational opportunities by enforcing on schools common performance standards for all inevitably leads to the neglect of children's personal and political liberties. This apparent inconsistency is, however, actually a consequence of the political choice to pursue equality of opportunity by means of a commitment to an unnecessarily broadly defined uniformity in children's school performance and a neglect of the role that diversity in student performance may have in fulfilling the requirements of social justice. As long as the goal of the public educational system is understood as producing a similarity in children's developed talents and abilities that has been determined by state political majorities, it is indeed inevitable that schools will neglect children's self-identified reasons for developing their potential—namely, their emerging personal and collective conceptions of the good—and the various cultural resources necessary for developing that potential. If, by contrast, schools were to understand their role as facilitating the development of children's conceptions of the good, personal and political, the common elements of children's education would be limited to what is necessary for their developing as their own persons and as thoughtful participants in the overlapping consensus, with a reasonable chance to cultivate their self-defined comparative advantages in a democratic political system and a market-based economic system.

One important lesson of this analysis is that social justice requires us to be mindful of the superstructure of the education system rather than, at least initially, of the specific details of the schools' curriculum and instructional procedures. Unless the purposes of that system are aligned with the principles of social justice in education, any adjustments in the details of schooling—such as whether the schools in Jamesville should emphasize teaching English as a second language or whether they should emphasize the traditional academic curriculum—are inevitably inconsistent with those principles. In effect, social justice requires us to widen our perspective on the educational system to determine whether its general structure aligns with our fundamental and shared political convictions about the way that the schools should contribute to the development of children in the context of their cultures and communities. In Jamesville, much of the difficulty lies in the misalignment of that framework. It is not that Jamesville's schools need simply less regulation by central authorities, as many of the local authorities and education professionals believe; rather, these schools need central regulation of a significantly different kind than is currently the case. The current accountability mechanism forces those local authorities to pursue a kind and degree of uniformity in children's school achievement that is incompatible not only with their development toward the personal and political liberties of the future adults they will become but also with fair chances to develop their potential to contribute to the fulfillment of their own and the society's aspirations. What is needed, then, is a reformulated framework of accountability that draws local authorities' attention and directs their efforts toward meeting the responsibilities they have to provide a genuinely liberating and democratic education, one in which all of their children will have reasonable opportunities to shape and to accomplish their personal and political purposes.

In part, this improved accountability system would require local authorities to develop a school curriculum that is inclusive of the liberty-consistent local cultures that represent for their children live possibilities for their conceptions of the good. In Jamesville, this requirement would mean that the school curriculum needs to take Latino cultures seriously as possibilities of this kind and not simply as obstacles that must be overcome for Latino children to gain access to the dominant Anglo culture. The details of this aspect of the curriculum depend on the configurations of Latino cultures in Jamesville, but those cultures should by means of the curriculum and instructional procedures be made equally available to all local children, not only to Latino children, which likely means that all children in the community should have access to the Spanish language in addition to English and to the significant ideas, practices, and artifacts in those cultures. Of course, we do not know enough about Jamesville or its region to specify in detail what other liberty-consistent cultures, religions, or ways of life to which this curriculum should also provide access because they represent live possibilities for children's conceptions of the good. In fact, the analysis of those matters is a primary responsibility of local education officials, and the accountability system should enable central authorities to determine whether Jamesville schools pursue this responsibility sincerely and sensitively. Moreover, in light of this local knowledge, the curriculum and instructional procedures in Jamesville should be developed to provide children an education in the common knowledge, skills, and attitudes necessary to gain access to these cultures; to make reasonable personal judgments about them; to become thoughtful participants in the local, state, and national overlapping political consensus; and to make judgments about their personal and political roles in the economy. Here, too, the accountability system should enable Jamesville to demonstrate and central authorities to evaluate how well the local curriculum recognizes these requirements and how well local pedagogy succeeds in enabling Jamesville's children to acquire this common learning.

Before continuing the analysis of the situation in Jamesville, we should note just how different and more robust this accountability system is in comparison with the so-called standards-based approach currently imposed by states and the federal government. For, on the one hand, a just accountability system is rightly concerned with the processes that schools utilize as well as the outcomes of schooling in that it examines, for example, the adequacy of local schools' procedures for determining the content of education in their specific cultural contexts and the effectiveness of the instructional methods adopted to make that content accessible. On the other hand, it defines broadly the outcomes with which it is concerned in that it cares about not only children's learning of specified academic skills as measured by standardized tests but also about children's ability to apply those skills to the personal and political judgments they make about and within the cultures to which they have access. In addition, that system of accountability is concerned with the attitudes that children demonstrate in making those applications, with whether, for example, they rely sensibly on the evidence available to them and take seriously the arguments that others make. It recognizes, in other words, that even common learning in schools can produce different results in children's lives and behavior when the cultural context of those schools is taken into account. Just how important the role of children's similar performance on standardized measures of learning in holding schools and their localities accountable for the purposes implied by social justice is not obvious because schools must attend to how children acquire that learning in their particular contexts and how they use it in the lives they are shaping for themselves and not to whether they can demonstrate their learning in an artificially standardized situation. In fact, as our analysis has already suggested, such similarity of performance in a standardized context is more likely to mean that schools are enforcing on children a uniform set of personal and political aspirations acceptable to a political majority rather than that they are enabling children to take responsibility for their own personhood and citizenship. For these reasons, the current system of accountability for student outcomes is inadequate for determining whether Jamesville's schools are meeting their responsibilities for both the procedures associated with and the results of their children's common learning.

This conclusion is very different from the claims of Jamesville's education officials and teachers that the system is unfair because it does not recognize how hard it is to produce uniform results in a diverse student population, results that we have found are likely to reflect only the values of the state's dominant culture or cultures. These officials and teachers do not question whether the production of those uniform results does justice to Jamesville's children and cultural communities; they simply assert that the expectation to produce such results is unfair to schools and teachers because it is difficult to satisfy.

The current system of accountability in Jamesville is particularly inadequate for determining whether localities are attending to their responsibilities to provide a meaningfully differentiated curriculum that enables children to develop their comparative advantages that result from variations in their emerging conceptions of the good. That system cannot help school officials, teachers, parents, or patrons ascertain whether the differentiated curriculum available in Jamesville, if any, adequately responds to the diverse objectives that children seek to pursue to realize their conceptions because it does not recognize that education may, in part, have appropriately different content for children with different educational and life goals. Second, it cannot help citizens assess whether children are given fair access to the various curricula because it does not attend to the processes of education, including the processes of student counseling and placement. Third, it cannot help determine whether children placed in the different curricula are provided with instruction that enables them to have roughly equal chances of success in pursuing their aspirations in comparison with others taking the same curriculum or other curricula because it does not measure outcomes that may differ from one child to the other. In fact, that system has nothing to say about whether and how the schools should provide curricular options once children have satisfied the minimum requirements of the uniform standards. Apparently, that system leaves these matters entirely to local prerogative, with no accountability whatsoever to central authorities for the justice of the decisions that localities make.

To some extent at least, the controversy over Jamesville's schools may reflect community dissatisfaction with either the availability of meaningful curricular options or the placement of children in those options. Both Latino and Anglo parents may have concerns related to the differentiated curriculum in Jamesville. For example, Latino parents' worries about discrimination may reflect, in part, the denial to their children of access to particular courses or programs in Carrington Middle School or the inadequacy of the preparation of Latino children for such options, and Anglo parents' claims about a decreased focus on academics may result from the reduced availability of specialized academic programs for their children. Of course, additional investigation of these parents' issues and the conditions in Jamesville's schools would be necessary to establish whether these possible sources of concern are actually in operation, whether, for example, the curricular options are indeed becoming limited or whether some children do not have a fair chance to enroll in them. We should note, however, that the uniform standards-based accountability system does not hold schools responsible for maintaining differentiated curricula or for providing equal access to them. Moreover, that system seems to give schools a reason to ignore parents' demands for adequately differentiated instruction and to limit its availability in favor of increased efforts to produce better and more uniform outcomes on standardized tests. Most important, however, socially just forms of curricular differentiation should respond to children's developing judgments about the ways of life that will prove meaningful to them, and not simply to parents' desires for forms of schooling that they prefer regardless of their children's judgments and experiences. As children's voices have not been heard except indirectly through their parents, no final determination can be reached about the justice of Jamesville's schools' opportunities for differentiation based on the information available thus far. In fact, it is the responsibility of local school officials to analyze and to understand children's aspirations as they plan for and implement curricular and instructional options. And it is the responsibility of central authorities to hold localities accountable for performing these complex intellectual and practical tasks effectively.

Therefore, what seemed to be a contradiction between the elements of social justice in education that focus on personal and political liberty and those that focus on equality of opportunity vanishes when one takes an appropriately broad perspective on the situation in Jamesville. From this perspective, providing Jamesville's children with equal educational opportunity depends fully on whether their education is preparing them to be their own persons and responsible citizens. For the substance of the opportunities that are to be equalized cannot be determined in the absence of detailed knowledge about the persons that children aspire to become. Moreover, an accountability system like the one operative in Jamesville violates the requirements of both liberty and equal opportunity in education because it, in effect, both imposes a particular conception of the good on the town's children and restricts the educational opportunities they have only to the learning that is compatible with that specific conception.

Finally, Jamesville's teachers are concerned about the adequacy of the funding available to meet the expectations articulated by the accountability system, and, as we have seen in the last chapter, the distribution of funding is, indeed, a concern of social justice in that the distribution should allow the various schools in the system an equal chance of meeting the social justice purposes of education.

That is, funding to accomplish those purposes should be distributed according to the extent that the characteristics of local children make them more or less expensive to educate. The recent substantial and increasing presence of Latino children in Jamesville's schools certainly implies that those schools should change their instructional programs to accommodate the cultures of the Latino community and the circumstances, talents, and aspirations of Latino children. At the very least, the limited English proficiency of Latino children suggests that the costs of these accommodations is likely to be higher than they were previously, and other characteristics of the town's Latino children about which we have no information, such as family poverty or cultural isolation, may possibly increase these costs as well. At the same time, however, characteristics of Jamesville's Anglo children—such as their unfamiliarity with or resistance to Latino cultures and the Spanish language—may also increase the costs of an education for personal and political liberty in this context. Thus, a reasonable case can be made that the influx of Latino children into Jamesville's schools imposes higher costs of educating both Latino and Anglo children. But this is a very different case than the one that Jamesville's teachers contemplate. Their rationale focuses entirely on the additional costs of teaching the dominant culture and the English language to Latino children and not at all on the costs of the mutual adjustments that all children must make so that they can recognize and respect the emerging conceptions of the good of all their schoolmates and the perspectives of all the members of their community.

Beyond these considerations, moreover, the case for increased funding in Jamesville depends on the costs of education that other communities in the state are facing as well. If the cultural or socioeconomic circumstances, for example, of other communities are more severe than they are in Jamesville, those communities may have an even stronger case for increased funding than Jamesville has. As we have seen, social justice in the distribution of funding for schools is a matter of vertical equity among all schools and school districts and not merely a matter of meeting the costs of education in any particular community. If Jamesville's teachers want to argue for more funding, they, first, need to assess the costs of meeting the socially just purposes of schooling in their community and, second, to compare these local costs with those of attaining social justice in other communities. The four principles, after all, require fairness across schools in the entire society.

In fact, this is the ultimate lesson of the analysis of the controversy over the schools in Jamesville: Justice in schooling necessitates a comprehensive perspective on education. Such a perspective requires, first, that all of the legitimate purposes of schooling included in the overlapping political consensus are considered, not, for example, just the society's interests in common learning to the exclusion of those in differentiated outcomes. Second, it requires a consideration of the aspirations of all children in a community, not just those favored by a political majority in the locality, the state, or the nation. Third, it requires a balanced consideration of the needs of all localities, not just what is needed by some particular locality. Thus, the result of the application of social justice to the education provided in Jamesville's schools does not necessarily align with the interests of a particular segment of that community—those of local officials, parents, teachers, or even children—or with interests of Jamesville as a whole.

As we have found, social justice calls for an adjustment in the framework that governs schooling in Jamesville, the framework that specifies the legitimate purposes of schools and the responsibilities for which the parties to that schooling, both local and central, should be held accountable. This revised framework and accountability system conflicts in some ways with the interests that have been expressed by all of those parties. It requires central authorities to regulate school outcomes, processes, and resources in a way that is at odds with their currently preferred (and much less demanding) standards-based approach. It requires local authorities to defer to central authorities regarding the purposes of schooling, however much they are tempted to want to control these purposes according to their own interests. It requires parents to honor the conceptions of the good that their children are developing even though they may not be entirely to their liking. It requires children to learn from and adjust to the conceptions of the good of other children and adults in the community and, to some extent, in the nation as whole. Moreover, it requires teachers to make do with a fair share of school resources and no more, even though having more than their fair share would make their work easier and quite possibly the school experience of the children with whom they work richer. In light of these observations, social justice in education is not a mechanism for making any of our or our children's fondest wishes come true, for that is and should rightly be the responsibility of each of us with the reasonable assistance of others in our families, communities, states, and the nation. In effect, the principles of social justice in education describe Americans' political consensus about just what sort of educational assistance is reasonable and therefore morally required for our children to have a fair chance of developing and pursing their own aspirations. A wider account of social justice, which this book does not attempt to articulate, would specify the assistance that is owed to members of our society by other sorts of social institutions. But in either case, the test of the social justice of our institutions is not whether they enable their citizens to succeed according to their self-identified aspirations, much less according to some externally prescribed and uniformly enforced conception of worthy human lives, but whether they have a fair chance of doing so.

Moreover, this framework does not specify or attempt to control the specific and concrete results of the school system. Rather, it establishes a superstructure for schooling in which the specific outcomes depend on judgments that the various parties to the system— central authorities, school officials and professionals, parents, other members of the community, and children—make in good faith, judgments that are likely to vary considerably from one community and one school to the other. In this way, social justice in education demands an institutional context within which the parties develop and exercise their freedom, mutual respect, political and moral responsibility, and intelligence.

In effect, social justice in the education of the young calls for schools to be sites for the moral, social, intellectual, and economic development of the communities in which they are situated. This is certainly the case for the children of those communities in that the human and material resources to be provided for their development have a bearing on all these qualities of children's character. In contributing to children's character in these ways, schools assuredly influence the future character of the community. But schools also can and should have an influence on the current character of their communities as well. After all, schools are places where the adult members of the community express their own deepest and most sincere beliefs about the lives that they find worth living in the form of the aspirations that they communicate for their children. In a real way, those expressions are more honest and less disguised by strategic considerations than they are certainly in business relationships or even in discussions of other, more distant functions of government. If those expressions are mediated and constrained by the requirements of the overlapping consensus about social justice in education, moreover, schools can be venues in which adults learn about other community members' conceptions of the good, discover and understand the justifications that those others have for their conceptions, and reach informed judgments about the reasonableness of others' conceptions and justifications. Of course, chances are that those judgments are not completely and inevitably sympathetic, but they are likely to reflect greater sincerity and toleration than those reached in other contexts where what is at stake is less important than the lives and futures of their own and others' children. While it is often maintained that schools are cradles of liberty and democracy, that usually is taken to mean only that schools are places where children learn to respect the personal and political rights of others. Although these implications clearly hold for children's learning on this account of social justice in education, the school, probably more than any other social institution, can thus also be the site of the mutual education of adults in the authenticity, open-mindedness, and forbearance that liberty and democracy require.

Notes

Preface

1. Some might argue that John Rawls (1999b) in his *Theory of Justice* and later works has accomplished or at least attempted this larger task. As I acknowledge in more detail later, Rawls is an important inspiration for the method I will use and the arguments I will make. However, Rawls's conclusions in his later works suggest that his method may lead to different results in different liberal democratic societies and even at different times within a particular society of that kind. For that reason, I focus my arguments on a particular nation and on a specific institution within that nation, namely education in the United States, about which there is plausibly a reasonable degree of public normative consensus on its operation and purposes.

Chapter Two   A Political Theory of Social Justice for Education: Liberty and Democracy

1. The ideas developed in this chapter and the next are an expansion and elaboration of those published in Bull (2007).
2. I recognize that in this chapter and elsewhere, I am taking considerable liberties with Rawls's idea of an overlapping political consensus. Here specifically, I am applying the idea to a particular institution rather than, as Rawls does, to the entire basic structure of society that is to govern the operation of all of society's most fundamental institutions. However, as Rawls himself acknowledges, the school is one of the fundamental institutions in the basic structure. Beyond this, as I have noted in the preface, this variation on Rawls's approach may be justified when a societywide consensus on the society's entire basic structure is not feasible but when there is enough apparent

agreement about particular institutions to determine what form social justice would take in those institutions. In the absence of a larger overlapping consensus, the society can at least seek social justice in the operation of particular institutions. Moreover, I do not place the principles of justice in education in a priority order for the purposes of resolving conflicts among them, as Rawls does. Rather, I attempt to formulate the principles in a way that minimizes those conflicts from the outset. In fact, there is evidence in Rawls's work (1999b) that he uses this approach to some extent in his formulation of principles of justice, when, for example, he limits the protected personal and political liberties to provide space for the other principles to operate. In future chapters, particularly chapter five, I note other major departures from Rawls's original understanding of the idea of an overlapping political consensus and provide a rationale for those alternative interpretations.

Chapter Three   A Political Theory of Social Justice for Education: Equal Opportunity and Economic Growth

1. I recognize that Gutmann conceptualizes her threshold specifically for the democratic theory of education that she develops. However, I have adapted her proposal to the theory being developed here, making appropriate adjustments where necessary. I briefly consider her specific interpretation of the threshold in chapter six.

Chapter Four   Personal Liberty and Education: Families, Cultures, and Standards

1. These three possibilities are not the only imaginable justifications of standards for schools, but only those relevant to the discussion provided earlier for an education for personal liberty, especially the requirement for access to multiple liberty-consistent cultures. However, these three justifications are of importance under this conception of social justice in education because the four principles were articulated to harmonize all four basic American political values about education. Thus, providing for children's future personal liberty is a requirement for an education also to meet the requirements of democracy, equal opportunity, and economic growth as they have

been conceptualized here, just as the attainment of equal opportunity, for example, is necessary for the realization of the other three principles.

2. This discussion is predicated on the definition that Hirsch provides of cultural literacy. In laying out the specific elements of cultural literacy, however, Hirsch seems to focus on the content of dominant cultures. Thus, his practice but not his theory of cultural literacy is culturally hegemonic.

3. Eamonn Callan (1996, 21) has argued that the politically liberal theory of justice that John Rawls develops and that is being developed here is equivalent to a partially comprehensive liberalism that he labels ethical liberalism and, thus, includes a commitment to what he calls "autonomy through the back door." He reaches this conclusion because of what he argues is the necessary form of civic education in a politically liberal society. I do not believe that this is so, but my reasons will have to await the next chapter in which I develop an account of the civic education that is implied by the principles of social justice in education.

Chapter Five   Democracy and Education: Multicultural and Civic Education

1. Thus, liberal political theories do not reflect the colloquial American meaning of "liberal" that is opposed to "conservative" because many, perhaps most, American liberals and conservatives place a central value on liberty.

2. It is not my intention to claim that my use of the idea of an overlapping consensus is precisely what Rawls intended by that term or that the subsequent analysis of civic education is necessarily one that Rawls would have endorsed. There are two elements in Rawls's own treatment of this idea from which my use of it varies. First, for Rawls the overlapping consensus is the third and final stage of a process by which a fully politically liberal society may emerge, one that follows a modus vivendi stage, in which citizens treat principles of political morality as instruments useful only to achieving their private ends, and then a constitutional consensus stage, in which citizens have a genuine but shallow commitment to some elements of liberal political morality as a framework within which they pursue their private ends (Rawls 1993, pp. 133–172). Thus, the overlapping consensus stage is

Rawls's technical term for citizens' coming to accept such a morality as fully regulative of their adoption and pursuit of their ends. I, by contrast, rely on an intuitive rather than this technical meaning of this idea. Second, Rawls in his 1971 *Theory of Justice* seemed to imply that the political morality of what he later called the overlapping consensus has a fixed substantive content derived logically from the premises on which it is based. However, by the time of *Political Liberalism* (1993) and *The Law of Peoples* (1999a), Rawls makes clear that the specific public principles that emerge from an overlapping consensus may take a wide variety of forms. Thus, he came to have a more procedural interpretation of an overlapping consensus, which is the meaning on which I rely and expand in this chapter.

3. I have reconsidered this issue since the time of my response to Kenneth Strike in Bull (1992).

Chapter Seven Economic Growth and Education: The Financing of Education

1. Of course, the economy is not the only social institution that performs such a social coordination function, for the political system and the private associations of civil society, for example, do so as well. However, the economy has an especially pervasive and important role even in the operation of these other institutions.

2. These observations about Brighouse's suggestion for economic education derive largely from two important features of his account of education that the overlapping consensus about education does not share. First, Brighouse posits a comprehensive theory of human value—namely, an account that establishes human flourishing as an ultimate value—which this book's political liberalism tries to avoid. Second and relatedly, Brighouse posits a general theory of social justice that goes beyond the theory in this book, which focuses exclusively on the overlapping consensus among Americans about education but not necessarily about society's many other institutions. As I suggested in the preface, it may be that such a general theory of justice will eventually turn out to be possible, but this book is focused only on the limited theory of justice for a single institution that the agreements among Americans about schooling seem to make possible at this moment.

3. The following account follows loosely that in Friedman (1962, 83–107).

4. I have expressed these problems, for the sake of simplicity in the exposition, as a threat of underinvestment in education, but the more accurate economic designation of these problems is as a threat of misallocation of education. In fact, uncertainty and wealth distribution can create overinvestment as well as underinvestment in education. Some optimistic individuals can calculate their predicted internal returns using figures higher than the average return, which can offset the decisions made by those who use lower amounts. Thus, the net investment effect of uncertainty depends on whether the population is overall risk averse. Similarly, some individuals from wealthy families can calculate their expenditures for education on the basis of what they can afford, which may be greater than the expected economic return. This, too, may offset the effects of the decisions made by low-income individuals. Here, the net investment effect of wealth distribution depends on the noneconomically determined education preferences of the wealthy and the poor. But in both cases, even if it so happens that individual overexpenditure exactly balances individual underexpenditure so that the theoretically optimal overall amount is spent on education, the outcome will be that the returns to the entire society are less than optimal because in individual cases underinvestment and overinvestment occur and result in a mismatch between those who receive education and those who have the potential to generate economic returns. Ignorance and the external effects of education do consistently lead to underinvestment, but these problems can also be understood to result in misallocation in that insufficient education is allocated to those who have the highest potential to create economic returns. This misallocation, too, generates returns to education that are less than economically optimal.

References

Abington School District v. Schempp. 1963. 374 US 203.

Ackerman, Bruce A. 1980. *Social justice in the liberal state.* New Haven, CT: Yale University Press.

Bellah, Robert N., Richard Madsen, William M. Sullivan, Ann Swidler, and Steven M. Tipton. 1991. *The good society.* New York: Alfred A. Knopf.

Brighouse, Harry. 2006. *On education.* New York: Routledge.

Brown v. Board of Education. 1954. 347 US 483.

Bull, Barry. 1984. Liberty and the new localism: Toward an evaluation of the tradeoff between educational equity and local control of schools. *Educational Theory* 34, no. 1: 75–94.

———. 1990. The limits of teacher professionalization. In *The moral dimensions of teaching,* edited by John Goodlad, Roger Soder, and Kenneth A. Sirotnik, 87–129. San Francisco: Jossey-Bass.

———. 1992. The creolization of liberalism. In *Philosophy of education, 1992,* Proceedings of the Forty-eighth Annual Meeting of the Philosophy of Education Society, 237–240. Champaign, IL: Philosophy of Education Society.

———. 1996. Is systemic reform in education morally justified? *Studies in Philosophy and Education* 15, no. 1: 13–23.

———. 2000a. National standards in local context: A philosophical and policy analysis. In *Educational leadership: Policy dimensions in the 21st century,* edited by Bruce A. Jones, 107–121. Stamford, CT: Ablex.

———. 2000b. Political philosophy and the balance between central and local control of schools. In *Balancing local control and state responsibility for K-12 education: 2000 yearbook of the American Educational Finance Association,* edited by Neil Theobald and Betty Malen, 21–46. Larchmont, NY: Eye on Education.

Bull, Barry. 2006a. Can civic and moral education be distinguished? In *Moral and civic learning in the United States,* edited by Donald Warren and John Patrick, 21–31. New York: Palgrave Macmillan.

Bull, Barry. 2006b. Is standards-based school reform consistent with schooling for personal liberty? *Studies in Philosophy and Education* 25, no. 1 & 2: 61–78.

———. 2007. A political theory of social justice in American schools. In *To what ends and by what means?: The social justice implications of contemporary school finance theory and policy,* edited by Gloria M. Rodriguez and R. Anthony Rolle, 9–34. New York: Routledge.

———. (In press). A politically liberal conception of civic education. *Studies in Philosophy and Education.*

Bull, Barry, Royal Fruehling, and Virgie Chattergy. 1992. *The ethics of multicultural and bilingual education.* New York: Teachers College Press.

Callan, Eamonn. 1996. Political liberalism and political education. *Review of Politics* 58, no. 1: 5–33.

———. 1997. *Creating citizens: Political education and liberal democracy.* New York: Clarendon Press.

Coleman, James S. 1966. *Equality of educational opportunity.* Washington, DC: U.S. Department of Health, Education, and Welfare.

Dahl, Robert A. 1971. *Polyarchy: Participation and opposition.* New Haven, CT: Yale University Press.

Dewey, John. 1927. *The public and its problems.* New York: H. Holt and Company.

Finn, Chester E., Jr., Diane Ravitch, and Robert T. Fancher, eds. 1984. *Against mediocrity: The humanities in America's high schools.* New York: Holmes & Meier.

Foley, Douglas A., Bradley A. Levinson, and Janise Hurtig. 2000. Anthropology goes inside: The new educational ethnography of ethnicity and gender. *Review of Research in Education* 25, no. 2000–2001: 37–98.

Friedman, Milton. 1962. *Capitalism and freedom.* Chicago: University of Chicago Press.

Fuhrman, Susan H., and Allan Odden. 2001. Introduction to the special section on school reform. *Phi Delta Kappan* 83, no. 1 (September): 59–61.

Gutmann, Amy (1999). *Democratic education.* Revised ed. Princeton, NJ: Princeton University Press.

Gutmann, Amy, and John Thompson. 1996. *Democracy and disagreement.* Cambridge, MA: Belknap Press.

Hess, Frederick M., and Michael J. Petrilli. 2007. *No child left behind: Primer.* New York: Peter Lang.

Hirsch, E. D., Jr. 1987. *Cultural literacy: What every American needs to know.* Boston: Houghton Mifflin.

———. 1996. *The schools we need and why we don't have them.* New York: Doubleday.

Howe, Kenneth R. 1997. *Understanding equal educational opportunity: Social justice, democracy, and schooling.* New York: Teachers College Press.

Kant, Immanuel. 1785/1985. *Foundations of the metaphysics of morals,* translated by Lewis White Beck. New York: Macmillan.

Lau v. Nichols. 1974. 414 US 563.

Macedo, Stephen. 1995. Liberal civic education and religious fundamentalism: The case of God v. John Rawls. *Ethics* 105, no. 3: 468–496.

———. 2000. *Diversity and distrust: Civic education in a multicultural democracy.* Cambridge, MA: Harvard University Press.

Mill, John. S. 1859/1978. *On liberty.* Indianapolis, IN: Hackett.

Mills v. Board of Education. 1972. 348 F Supp 866.

Mozert v. Hawkins County Board of Education. 1987. 827 F.2d 1058 (6th Cir.).

Noddings, Nel. 2003. *Caring: A feminine approach to ethics & moral education.* 2nd ed. Berkeley, CA: University of California Press.

Nozick, Robert. 1974. *Anarchy, state, and utopia.* New York: Basic Books.

Nussbaum, Martha C. 2006. *Frontiers of justice: Disability, nationality, species membership.* Cambridge, MA: Harvard University Press.

Pierce v. Society of Sisters. 1925. 268 US 510.

Ravitch, Diane. 2000. *Left back: A century of failed school reforms.* New York: Simon & Schuster.

Rawls, John. 1993. *Political liberalism.* New York: Columbia University Press.

———. 1999a. *The law of peoples.* Cambridge, MA: Harvard University Press.

———. 1999b. *A theory of justice.* Revised ed. Cambridge, MA: Harvard University Press.

Sandel, Michael, 1996. *Democracy's discontent: America in search of a public philosophy.* Cambridge, MA: Harvard University Press.

Sen, Amartya. 2002. *Rationality and freedom.* Cambridge, MA: Harvard University Press.

Strike, Kenneth A. 1994. On the construction of public speech: Pluralism and public reason. *Educational Theory* 44, no. 1: 1–26.

Strike, Kenneth A. 2004. Community, the missing element of school reform: Why schools should be more like congregations than banks. *American Journal of Education* 110, no. 3: 215–232.

Taylor, Charles. 1989. *Sources of the self: The making of the modern identity.* Cambridge, MA: Harvard University Press.

Thurow, Lester C. 1970. *Investment in human capital.* Belmont, CA: Wadsworth.

Tinker v. Des Moines Independent Community School District. 1969. 393 US 503.

Todd v. Rochester Community Schools. 1972. 200 NW2d 90.

Tucker, Marc S., and Judy B. Codding. (1998). *Standards for our schools: How to set them, measure them, and reach them.* San Francisco: Jossey-Bass.

U.S. Congress. 2001. *No child left behind act.* Public Law 107–110.http://frwebgate.access.gpo.gov/cgi-bin/getdoc.cgi?dbname=107_cong_public_laws&docid=f:publ110.107.pdf (accessed February 8, 2007).

INDEX